Future Developments in the Genetic Improvement of Animals

Contributors

Robert D. Anderson
I. R. Franklin
I. K. Gauld
Keith Hammond
J. W. James
L. P. Jones
R. B. Land
G. J. Lee
David Lindsay
A. E. McClintock
Harald Skjervold
M. J. Stear
J. F. Taylor
K. A. Ward
R. Webb

Future Developments in the Genetic Improvement of Animals

Edited by

J. S. F. Barker
Department of Animal Science
University of New England
Armidale N.S.W.

Keith Hammond
Animal Genetics and Breeding Unit
University of New England
Armidale N.S.W.

A. E. McClintock
Animal Genetics and Breeding Unit
University of New England
Armidale N.S.W.

1982

ACADEMIC PRESS
A Subsidiary of Harcourt Brace Jovanovich, Publishers

SYDNEY NEW YORK LONDON
PARIS SAN DIEGO SAN FRANCISCO SAO PAULO TOKYO TORONTO

ACADEMIC PRESS AUSTRALIA
Centrecourt, 25–27 Paul Street North
North Ryde, N.S.W. 2113

United States Edition published by
ACADEMIC PRESS INC.
111 Fifth Avenue
New York, New York 10003

United Kingdom Edition published by
ACADEMIC PRESS, INC. (LONDON) LTD.
24/28 Oval Road, London NW1 7DX

Printed in Australia

National Library of Australia Cataloguing-in-Publication Data

Future developments in the genetic improvement of animals.

Bibliography.
Includes index.
ISBN 0 12 078830 6.

1. Breeding. 2. Genetics. I. Barker, J. S. F. (James Stuart Flinton), 1931-. II. Hammond, Keith, 1942- . III. McClintock, A. E. (Alexander E.).

636.08'21

Library of Congress Catalog Card Number: 82-71824

Academic Press Rapid Manuscript Reproduction

Contents

Part 3 Immunogenetics

Part 4 Reproductive Biology

Part 5 Economic Aspects of Developing Breeding Objectives

Part 7 Population Size

Part 8 Electronics

Contributors

Numbers in parentheses indicate the pages on which the authors' contributions begin.

Anderson, Robert D. (141), Department of Sheep Husbandry, Massey University, Palmerston North, New Zealand.

Franklin, I. R. (181), Genetics Research Laboratories, CSIRO, P.O. Box 184, North Ryde, N.S.W. 2113, Australia.

Gauld, I. K. (59), Animal Breeding Research Organisation, West Mains Road, Edinburgh, EH9 3JQ, U.K.

Hammond, Keith (197, 215), Animal Genetics and Breeding Unit, University of New England, Armidale, N.S.W. 2351, Australia.

James, J. W. (107), School of Wool and Pastoral Sciences, University of New South Wales, P.O. Box 1, Kensington, N.S.W. 2033, Australia.

Jones, L. P. (119), Animal Research Institute, Werribee, Vic. 3030, Australia.

Land, R. B. (59), Animal Breeding Research Organisation, West Mains Road, Edinburgh, EH9 3JQ, U.K.

Lee, G. J. (59), Animal Breeding Research Organisation, West Mains Road, Edinburgh, EH9 3JQ, U.K.

Lindsay, David (89), Department of Animal Science and Production, The University of Western Australia, Nedlands, W.A. 6009, Australia.

McClintock, A. E. (157, 215), Animal Genetics and Breeding Unit, University of New England, Armidale, N.S.W. 2351, Australia.

Skjervold, Harald (3), Agricultural University of Norway, Institute of Animal Genetics and Breeding, Box 24, 1432 Ås-NLH, Norway.

Stear, M. J. (45), Department of Immunology, John Curtin School of Medical Research, Australian National University, Canberra, A.C.T. 2601, Australia.

Taylor, J. F. (157), Animal Genetics and Breeding Unit, University of New England, Armidale, N.S.W. 2351, Australia.

Ward, K. A. (17), Division of Animal Production, CSIRO, Ian Clunies Ross Animal Research Laboratory, P.O. Box 239, Blacktown, N.S.W. 2148, Australia.

Webb, R. (59), Animal Breeding Research Organisation, West Mains Road, Edinburgh, EH9 3JQ, U.K.

Invited Symposium Participants

Barker, J. S. F., Department of Animal Science, University of New England, Armidale, N.S.W. 2351, Australia.

Bindon, B. M., CSIRO, Private Mail Bag, Armidale, N.S.W. 2350, Australia.

Frankham, R., School of Biological Sciences, Macquarie University, North Ryde, N.S.W. 2113, Australia.

Goddard, M. E., Division of Animal Production, Department of Tropical Veterinary Science, James Cook University, Townsville, Qld. 4811, Australia.

Graser, H-U., Animal Genetics and Breeding Unit, University of New England, Armidale, N.S.W. 2351, Australia.

Hancock, T. W., Waite Research Institute, University of Adelaide, Adelaide, S.A. 5000, Australia.

Hetzel, D. J. S., CSIRO, Private Mail Bag, Armidale, N.S.W. 2350, Australia.

Lewer, R., Animal Research Institute, Department of Agriculture, Katanning District Office, Katanning, W.A. 6317, Australia.

McPhee, C. P., Animal Research Institute, Yeerongpilly, Qld. 4105, Australia.

Piper, L. R., CSIRO, Division of Animal Production, Chiswick, N.S.W. 2350, Australia.

Rathie, K. A. Dairy Cattle Husbandry Branch, Department of Primary Industries, Brisbane, Qld. 4001, Australia.

Robinson, G., Division of Dairying, Department of Agriculture, Box 4041, G.P.O., Melbourne, Vic. 3001, Australia.

Sheldon, B. L., CSIRO, Genetics Research Laboratories, Division of Animal Production, P.O. Box 184, North Ryde, N.S.W. 2113, Australia.

Sheridan, A. K., Poultry Research Station, P.O. Box 11, Seven Hills, N.S.W. 2147, Australia.

Wickham, B. W., Farm Production Division, New Zealand Dairy Board, P.O. Box 417, Wellington, New Zealand.

Yoo, B. H., CSIRO, Genetics Research Laboratories, Division of Animal Production, P.O. Box 184, North Ryde, N.S.W. 2113, Australia.

Preface

The intensive application of genetics in animal breeding dates from only about 40 years ago. Over the last 25 years, there has been increasing application of quantitative genetic theory in animal breeding programs. This has been facilitated by:

1 experimental validation of that theory and the translation of the theory in the development of appropriate operational procedures for practical animal breeding programs,

2 objective measurement of animal performance, the use of computers in large-scale performance recording programs, the application of artificial insemination in and the development of national or regional dairy cattle breeding programs, and the recognition by the animal industries of the cumulative gains possible from genetic improvement.

At the same time, the science of genetics has developed enormously, and expanded and divided into specialized areas including molecular genetics, immunogenetics, quantitative (or statistical) and population genetics. All of these have implications for and applications in the economic improvement of domestic livestock.

Given the relatively short time-span, the contributions of quantitative genetics to improvement in animal production have been substantial. However, it is apparent that future improvements depend not only on this branch of the science. It seemed appropriate then to bring together ideas from various branches of genetics and to consider them, together with recent developments in reproductive biology and physiology and electronics, so as to provide a perspective of their interrelationships and their possible future development and application in the genetic improvement of animals.

This book consists of the invited papers presented by scientists from New Zealand, Norway, U. K. and Australia, and summaries of discussions at a Symposium held at the University of New England, Armidale in February 1982. The objective of the Symposium was to project, from current knowledge, those areas where the most important developments in the application of genetics and associated disciplines could and should take place. The book thus represents a step towards the necessary integration of these branches of genetics and related disciplines.

The Symposium would not have been possible without the generous financial support of the Australian Meat Research Committee and The Angus Society of

Australia. We wish to thank the contributors for their cooperation and excellent papers, and to acknowledge the enthusiasm, views and ideas of all participants in the Symposium. Our thanks also go to Dorothy Cordingley and Jill Parker for their untiring assistance with the preparation of the final manuscripts and index.

Part 1
A General View of Animal Breeding

The Results of 20 Years Selection for Production in Cattle, Sheep and Pigs — Which Way Now?

Harald Skjervold

Department of Animal Genetics and Breeding
Agricultural University of Norway
Norway

Research, development and implementation of animal breeding techniques over the past 20 years have given good results. The rate of genetic improvement has increased considerably in all farm species.

Despite this success, optimum rates of improvement are, for different reasons, not yet being realised.

The decisions for increased genetic improvement in the later 1980's are now being implemented, but a biological delay precedes the expression of this.

Technical innovations are being researched and developed for use in increasing genetic improvement. Biophysical profiling, embryo manipulation, genetic engineering, computer tomography and multibreed mating strategies are examples of techniques which may become commonplace before the end of the century.

I A Review of the Past 25 Years

Twenty-five years ago, in the 50th anniversary issue of the Journal of Dairy Science, Jay L. Lush reviewed the improvement which had been made in animal breeding during the previous few decades.

Dr. Lush mentioned the development of biometrical methods, the evaluation of different selection methods, and the

FUTURE DEVELOPMENTS IN THE
GENETIC IMPROVEMENT OF ANIMALS
ISBN 0 12 078830 6

introduction of artificial insemination. He remarked: "Some of the parameters are known only within wide limits, and much is uncertain ... but at least a good beginning has been made."

During the last 25 years, animal genetics and breeding has been very productive. The purpose of this address is to review the major components that have contributed to this great progress in animal breeding. The intention is also to venture some predictions.

A. Developmental Work

During the last two decades, we have witnessed accelerated activity within the field of quantitative genetics. This development is especially pronounced in investigations which are of fundamental interest in the applied field of this science, such as:

(1) Estimation of phenotypic and genetic parameters. As shown in Figure 1, this developmental work exhibits an exponential growth. We know these parameters for an increasing number of traits and for an increasing number of populations and the estimations are based on steadily increasing amounts of data.

(2) The last two decades are also characterised by great improvements in the accuracy of estimation of breeding values. Progress is notable in methods of progeny testing bulls, boars and rams. The key-word in this connection is BLUP (Best Linear Unbiassed Prediction).

(3) Much work has been done in developing effective selection indices which can be directly applied in practice.

(4) Since the late 50's, the development of integrated breeding schemes has begun. This synthesis of different elements within quantitative genetics has gradually resulted in more efficient breeding strategies being put into use.

B. The Computer

The computer revolution has passed through a handful of "generations" of equipment in the past 20 years, and the rate of change does not seem to decrease.

Thanks to the development of the computer, it has been possible to conduct the enormous quantity of developmental work mentioned above. Without the use of computers, it seems unlikely that quantitative genetics really could have been introduced into animal breeding in the way it has been.

Without computers and with the cost of labour today, it would be impossible to estimate all the parameters and calculate the selection indices, etc. on which animal breeding is based today.

The computer has made it possible to use sophisticated statistical models and to store and utilise all relevant information when evaluating breeding values. Sophisticated evaluations of sires and dams are now possible using BLUP. Thanks to the computer, it has also been possible to practise selection among females based on breeding indices (cow index, sow index and ewe index). Further, all relevant data for each female can be stored in a data bank to have an up-to-date "account" for each animal.

Apart from packaged statistical programs, many other opportunities for aiding research have come about because of progress in computer science. Various aspects of simulation and systems analysis have been useful in different branches of animal science. Such simulations have been very useful in the development of integrated breeding schemes.

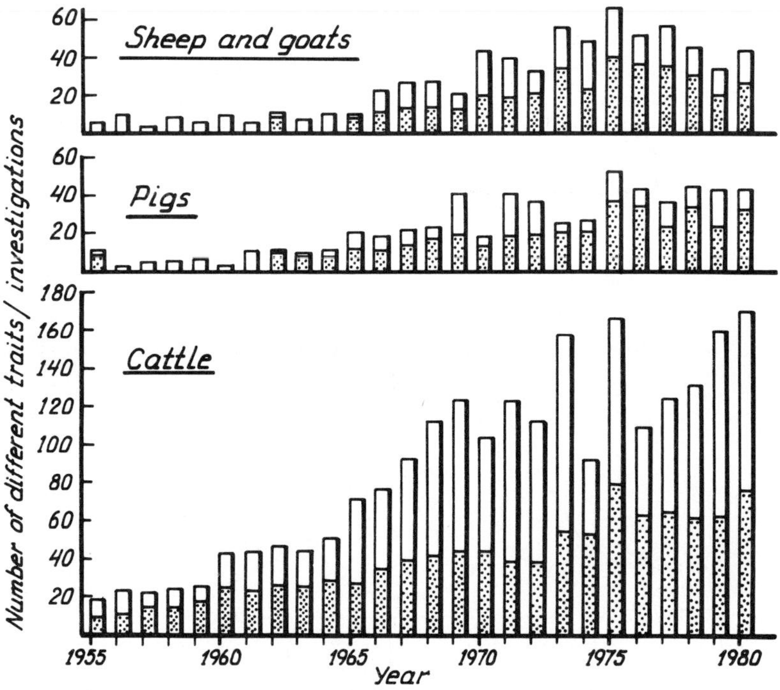

FIGURE 1. Number of publications concerning estimation of heritability 1955-80.

C. The Development of Artificial Insemination

Parallel to the development of applied quantitative genetics we have witnessed a large expansion in AI during the 60's and 70's. Twenty years ago AI was in its "teens" in most countries. During the last two decades AI has become the most important mating method and, moreover, it has given cattle breeding a new dimension (Fig. 2). Its importance in cattle breeding has been further emphasised by the introduction of deep frozen semen.

During the last 20 years AI also has been introduced into pig breeding and in some countries more than half of the services are covered by AI. However, this method has not yet contributed very much to genetic improvement as cryopreservation of semen has not been successful in the pig.

D. Recording Schemes

Recording schemes, which greatly influence the results of animal breeding, have shown positive developments during the last 20 years.

The proportion of cows included in milk recording schemes has increased in most countries. This is very interesting especially when noting the increase in labour costs which

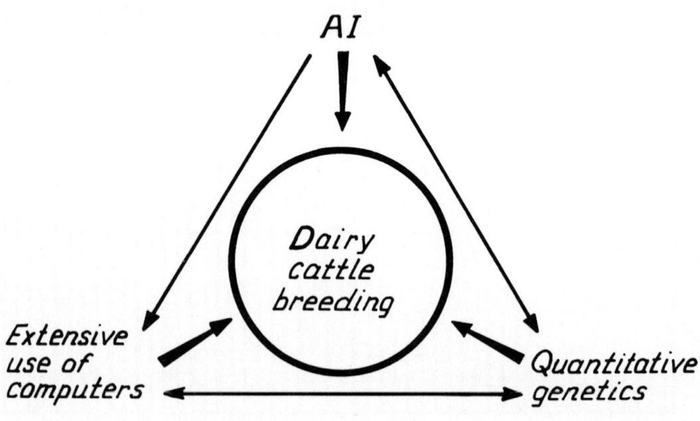

FIGURE 2. Dairy cattle breeding today is a synthesis of artificial insemination, quantitative genetics and extensive use of computers.

have occurred during these years. There may be various explanations for this trend, but there is no doubt that the introduction of the computer is an important reason.

The computer has reduced the work connected with the bookkeeping, at the same time it has expanded the use of the collected data through application of more efficient statistical techniques, greatly increasing our ability to scan for underlying trends and effects.

Today, records can be used more efficiently in their application to selection, dietary planning, economic analysis, etc. During recent years, we have also seen the beginning of an integration of cow health information with regular milk recording.

Sow recording schemes are now being developed in many countries in Europe. Information about litter size and performance is recorded.

Sheep recording schemes have also been started in some West European countries. Here, data are collected in order to facilitate progeny testing of rams.

E. Testing Schemes

The last 20 years also have been characterised by an increasing use of testing schemes, namely:

(1) Performance testing for growth rate, feed efficiency, etc. of young AI bulls, and of young bulls used in natural service.

(2) Performance testing of young boars (rate of gain, feed efficiency, backfat thickness, etc.).

(3) Sib-testing for carcass characteristics in pig breeding.

(4) "On-the-farm" performance testing of young gilts and boars.

F. Genetic Improvement

Great change in 'know-how' and resources has taken place within animal breeding during the last 20 years. This conclusion raises the question: *Has this effort resulted in increased genetic improvement?*

An important question from a practical viewpoint is how well the achieved genetic gain corresponds with the expected gain. It is not possible to mention all investigations here. To summarise:

(1) Most predictions indicate a theoretical gain which is significant. The expected gain is, however, in most cases considerably less than the maximum gain (in cattle) mentioned by Robertson and Rendel (1950).

(2) Analyses of expected genetic selection differentials along each of the four paths:

- females to produce daughters,
- females to produce sons,
- sires to produce daughters, and
- sires to produce sons,

have usually demonstrated the reasons why the achieved genetic gain is less than the predicted maximum gain.

In dairy cattle breeding the main reasons for the disagreement seem to be:

(1) Too low a selection intensity among bull sires.
(2) Too low a selection intensity among proven bulls.
(3) Too long a generation interval for bull dams.
(4) Too much attention paid to conformation.

In some of the pig data the theoretical gain for meat characteristics is often in the neighbourhood of what could be expected. However, the gain for litter size indicates a very low selection intensity. Data from sow recording in Norway, Poland and Sweden, and data from breeding herds in Denmark, indicate a selection intensity for litter size which is even less than natural selection for litter size[1]. This is because of its negative genetic correlations with some other important traits.

Some of the cattle populations which have used an integrated breeding scheme since the beginning of the 60's show genetic gains as high as 1% per year. In most populations, however, the trend for genetic gain is less than 0.5%. This is because selection in practice places considerable emphasis on non-productive traits.

Even though the genetic change in most populations is slower than what should be possible, there is good reason to believe that during the last two decades the rate of genetic

[1]*Natural selection for litter size according to Dickerson, is given by* $i = \sigma^2/\overline{X}$, *where* σ^2 *is the variance of litter size and* $\overline{X}$ *the average litter size.*

improvement has increased. Because of the great timelags experienced in animal breeding we have not yet reached the high rates of improvement expected when an integrated breeding scheme is fully underway. The conservative attitudes in the world of animal breeding mean that it often takes more time to change breeding strategies than is usually assumed.

II Future Developments

Of course, in years to come there is still much development work to do in order to improve efficiency. However, without gaining important *new knowledge*, it does not seem likely that we will be able to really lift the "ceiling" for possible genetic gain (Fig. 3).

Where do we expect to find such new knowledge? A common problem in animal breeding is the low heritability of many important traits. As pointed out by Land *et al.* (1981), the low heritability of reproduction traits does not really arise from low genetic variance, but rather from large environmental variation. The coefficient of variation is of the order of 40% whereas traits of high heritability such as growth

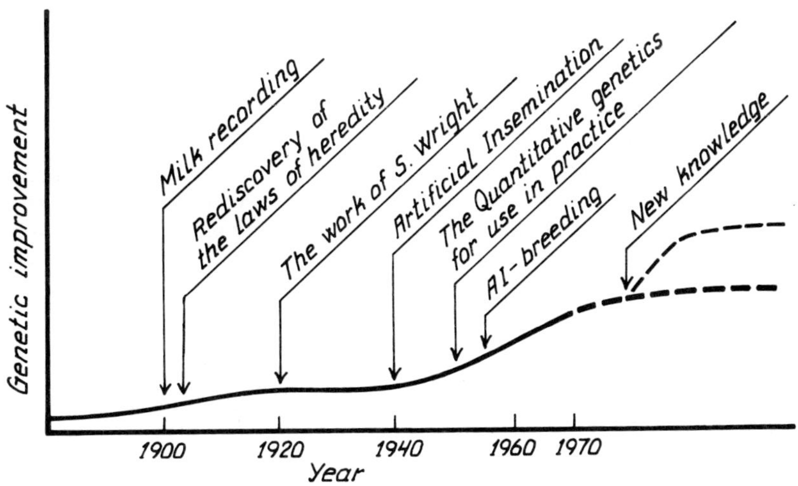

FIGURE 3. Factors contributing to the rates of genetic improvement attainable today. Any further increase requires new technology.

rate have coefficients of variation of about 10%. The potential for genetic gain is therefore as great as for traits having high heritabilities. The problem in these circumstances is how to identify quickly and accurately which are the genetically superior animals, and with minimum cost. Similar problems beset the improvement of milk yield and numerous other quantitative traits. Therefore there is considerable potential to increase the opportunities for genetic improvement through *better identification of genetic merit*.

Different approaches to achieve this are possible, namely:

(1) A selection strategy which, in addition to the trait in question, is also based on marker traits (indirect selection).

(2) More accurate performance recording of actual and "new" traits.

(3) Introduction of new bio-techniques.

A. Marker Traits

Let us start with indirect selection based on marker traits. These traits might be enzymes, hormones, different proteins, blood groups, etc. Some of these traits are easy to identify, but some of them require extensive biological tests for identification or quantification.

During recent years many investigations have been carried out to study the possibility of using marker traits when breeding for disease resistance. Most of the available information about the genetics of the immune response and correlated resistance traits is based on selective breeding in mice (for a review, see Biozzi *et al.*, 1979). An important observation in these studies in mice was that genetic regulation of the quantitative immune response was essentially nonspecific with regard to the antigen injected. The high responder animals were high responders to most of the complex antigens tested, as well as to antigens of pathogenic microorganisms.

Genetic variation of quantitative immune response has been demonstrated in bulls (Lie, 1979), pigs (Buschmann, 1980; Huang, 1977), goats (Almlid *et al.*, 1979) and chickens (Zijpp *et al.*, 1980). So far very limited information exists about genetic correlations between immune response data and resistance to infections in domestic animals. However, preliminary studies in cattle suggest that an association exists between antibody response in AI bulls and the incidence of clinical mastitis in their half-sibs in dairy herds (Lie and Solbu, 1981).

Our present knowledge about the importance of different resistance traits is of course too small to make detailed suggestions about the use of marker traits in selection for improved resistance to infections. However, it seems reasonable that in years to come we will introduce into the performance testing of bulls and boars, tests for marker traits associated with resistance to important diseases in cattle, goats and pigs.

As mentioned by Broster *et al.* (1969), a principal component of variation in milk producing ability may be the variation in the energy supply to the gland. Therefore, if one characteristic of a good cow is the ability to cope when suffering an energy deficit, the induction of a negative energy-balance in male calves by fasting might reveal relevant physiological covariation. Preliminary results indicate that this may be the case. In fact, *free fatty acids* increased more rapidly and urea-nitrogen more slowly when Friesian calves of high genetic merit were fasted, as compared with fasted calves of low merit (Tilakaratne *et al.*, 1980). Investigations by Joakimsen *et al.* (1971) show significant correlations between thyroid hormone activity in bull calves and the average milk yield of their daughters.

These results indicate that genes which influence the desired trait of the adult female may be expressed in the young male. Similar conclusions were drawn by Land *et al.* (1981), who found greater testis growth in males from lines in which female fertility was high, as compared with males from low female fertility lines.

B. In Vivo *Estimation of Body Composition*

A more accurate estimation of *in vivo* body composition would be of great interest in both animal breeding and nutrition science. In animal production the output of edible material per unit input of feed (*feed efficiency*) is an important component of the breeding aim. Genetic improvement of this trait is, however, limited because of the difficulty in obtaining an accurate *in vivo* estimate of body composition. Moreover, the lack of such estimation reduces the genetic improvement of carcass quality.

The introduction of ultrasonic equipment in the late fifties represented an important step in *in vivo* estimation of backfat thickness. However, the accuracy of estimates of body composition using this method is limited.

The introduction of computerised tomography seems to represent a break-through in *in vivo* estimation of body composition. Computerised tomography (CT) means the presentation of anatomical information from a cross-sectional plane of the body by computer synthesis of an image from X-ray transmission data. For the development of this technique, A.M. Cormack and G.N. Houndsfield received the Nobel Prize in 1979.

The CT equipment, which was developed for X-ray diagnosis in human medicine, also can be used for estimation of energy content and body composition on live pigs and sheep as shown by Skjervold *et al.* (1981). Their *in vivo* estimations were extremely good as compared with results obtained by other known methods. However, even greater accuracy is expected by combining information from several tomographic planes.

With this technique it seems possible to measure *in vivo* body composition with the accuracy which is desired when selecting breeding animals. In station testing of potential breeding animals, where carcass traits are involved in the breeding goal, the use of CT would largely remove the need for sib-tests to record carcass characteristics. Future carcass grading at abattoirs is another possible use of the body scanner.

In years to come new techniques will be developed in this field, which may be of great importance in animal science. In addition to the Computer Tomograph, the arrival of the Positron-Emission Tomograph and Dynamic Spatial Reconstructors have already been announced.

The rapid development within the electronics industry and of similar techniques to the tomograph is expected to produce spin-offs of importance for the further development of animal breeding. It is, however, dangerous to forecast what type of new technical aids will be of greatest importance.

C. Other Innovations

As we all know, there is at the moment great development activity in bio-technology and recently also in genetic engineering. It is expected that this extensive research work will contribute positively to future animal breeding. I shall only outline some of the most promising topics, namely:

1. Embryo Transplantation. The effect of this on livestock improvement includes:

(1) Embryos can be frozen and stored for use in future breeding programs (measurement of genetic drift).

(2) Selection intensity on females can be increased.

(3) The "grading up" process can be accelerated.
(4) The evaluation of foreign breeds is assisted.
(5) The twinning rate in cattle can be increased.

2. In Vitro *Manipulation of Embryos.* This technique belongs perhaps to the more distant future. However, it opens up a wide range of possible developments, including:

(1) Sexing of embryos,
(2) Nuclear transplantation,
(3) Splitting embryos,
(4) Cloning.

3. Development of Synthetic Breeds. So far we have been dealing with questions concerning additive genetic improvement. What about the breeding methods? Do we expect any change in the mating strategies?

In years to come, pure breeding will, without doubt, continue to be the common breeding method. We may, however, expect some increase in crossbreeding and not only within regular crossbreeding schemes. The development of synthetic breeds seems to have some important advantages and the synthetic seems to have an interesting future, especially in cattle.

REFERENCES

Almlid, T., Steine, T., Larsen, H.J., and Lunde, A. (1979). *30th Ann. Meeting, EAAP, Harrogate*, 1979.

Biozzi, G., Mouton, D., Sant'Anna, O.A., Passos, H.C., Gennari, M., Reis, M.H., Ferreira, V.C.A., Heumann, A.M., Bouthillier, Y., Ibanez, O.H., Stiffel, C., and Siqueira, M. (1979). *Curr. Top. Microbiol. Immunol. 85*, 31.

Broster, W.H., Broster, V.J., and Smith, T. (1969). *J. agric. Sci., Camb. 72*, 229.

Buschmann, H. (1980). *Proc. XVIIth Conf. on Animal Blood Groups and Biochemical Polymorphisms, Wageningen*, July/August, 1980.

Huang, J. (1977). *Dissertation, University of Hawaii, University Microfilms International.* Ann Arbor, Michigan.

Joakimsen, Ø, Steenberg, K., Lien, H., and Theodorsen, L. (1971). *Acta Agric. scand. 21*, 121.

Land, R.B., Baird, D.T., and Carr, W.R. (1981). *J. Reprod. Fert. 62*, 151.

Lie, Ø. (1979). *Acta vet. scand. 20*, 372.

Lie, Ø., and Solbu, H. (1981). *Acta vet. scand.* (in press).
Robertson, A., and Rendel, J.M. (1950). *J. Genet. 50*, 21.
Skjervold, H., Grønseth, K., Vangen, O., and Evensen, A. (1981). *Z. Tierzücht. ZüchtBiol. 98*, 77.
Tilakaratne, N., Alliston, J.C., Carr, W.R., Land, R.B., and Osmond, T.J. (1980). *Anim. Prod. 30*, 327.
Zijpp, A.J. van der, and Leenstra, F.R. (1980). *Poult. Sci. 59*, 1363.

Part 2
Molecular Genetics

Possible Contributions of Molecular Genetics to Animal Improvement

K. A. Ward

CSIRO
Division of Animal Production
Blacktown, NSW

Rapid increases in the understanding of the eukaryotic genome have recently taken place as a direct result of the discoveries of a number of revolutionary techniques which have made it possible to isolate and characterize individual genes. The new knowledge is certain to have a profound impact on the techniques used for the genetic improvement of our domestic breeds of animals.

In this paper the current status of our knowledge of eukaryotic gene structure is briefly reviewed, together with the techniques of recombinant DNA and gene transfer. The possible areas of application of these techniques are examined using the wool keratin gene complex as an example of a group of genes which should be readily amenable to manipulation. The difficulties imposed by the involvement of multiple gene action on a desired animal phenotype are discussed. In addition, the potential application of the new technology to rumen function and to conventional animal breeding are considered.

I Introduction

The improvement in the genetic base of our domestic animals has been the goal of most livestock management programmes since the time *Homo sapiens* first forsook the role of hunter/gatherer and assumed the tasks of farming. During

FUTURE DEVELOPMENTS IN THE
GENETIC IMPROVEMENT OF ANIMALS
ISBN 0 12 078830 6

this period, which began some 10,000 years ago, vast gains have been made in the selection of animals for specific purposes, with the results we see around us today. In fact, so well-adapted are most of our domestic livestock for their specific roles in the domestic scene of today, that we may well ask ourselves, to what extent can significant improvements in performance be expected in the future, and what role will genetics assume in any such improvements? In order to answer these questions, we need first to consider the traditional method which has given rise to genetic gains. In its simplest form, this involves choosing an animal with a desired characteristic and using this animal as a base for future breeding. A refinement of this approach is the identification of a desired characteristic in a breed which, however, also contains a number of undesirable characteristics, the mating of this breed with another breed which does not contain the undesirable characteristics, and the selection of desired progeny. However, the method is immensely complicated by the re-organisation during each breeding cycle of the large number of genes which make up the genetic background of the animal. While these genes may not be directly associated with the specific performance characteristic of interest, nevertheless new combinations can influence its expression in indirect ways. The prediction of such interactions is obviously very difficult. Elegant techniques have been developed over the years to at least partially predict and overcome these problems, and much will be said of these in other sections of this symposium.

Clearly, many of the above difficulties would be circumvented if the specific genes involved in the desired characteristic could be transferred from one animal to another without having to involve the general gene reshuffling which occurs in normal mating. It is in this area that molecular genetics may well make a major impact in the near future.

Molecular genetics and animal breeding have traditionally been somewhat removed from one another. Although the knowledge gained from molecular genetics has provided a base for the establishment of many of the genetic principles which have been applied to animal breeding, the enormous genetic diversity of an animal has proved to be a major stumbling block in the analysis of most genetic complexes involved in useful production characters. However, molecular genetics is currently undergoing an expansion in knowledge which can be described as explosive. Single genes can be readily isolated from any organism and analysed molecule by molecule. Furthermore, isolated genes can be directly transferred from one animal to another, and most importantly, they appear to be under control of the new host's cellular control mechanisms (Constantini and Lacy, 1981; Wagner *et al.*, 1981; Brinster *et al.*, 1981). The full impact of this new technology on the future is hard to predict, but it is clear that we are entering a new era in biology, and that entirely new concepts in animal breeding will shortly be upon us. In this paper the major events leading to the exciting situation which currently exists in this area, and the potential for the future application of the technology will be reviewed and discussed. By way of example, I will make specific reference at times to the wool keratin genes both because our knowledge of these is well-advanced and because they are of considerable importance in animal production.

II Recent Developments in Molecular Biology

The major discoveries in molecular genetics were initially made in bacteria and lower eukaryotes. During this period, which extended up to about the beginning of the last decade, attempts to extend the findings to higher eukaryotes met with only limited success, for several reasons. The techniques of investigation were aided by the short generation times of bacteria and the ability of the organism to grow under carefully defined growth conditions. In addition, the complexity of the genetic information in bacteria is several orders of magnitude less than that of the higher eukaryotes. Finally, the higher eukaryotes are under the control of a very strict programme during early development, so that any experimental manipulation which causes a significant deviation from the programme is usually lethal.

Within the last ten years this situation has altered dramatically, and we are now in the position to investigate the genetics of the higher eukaryotes at the same level as we were previously able to do in bacteria. This recent accelerated pace of discovery is due to the development of three important techniques.

The first of these is the improvement in methods for the growth of cells in culture, which now allow the majority of eukaryotic cells to be grown in large quantity in at least semi-defined media, and in many cases these cells can be experimentally induced to differentiate and express specific genetic properties. Thus one of the big stumbling blocks to the extension of bacterial genetic techniques to eukaryotes has been largely overcome.

Secondly, the complexity of the DNA of higher organisms as compared with bacteria has been overcome by the finding of specific endonucleases called restriction enzymes. These enzymes, which are generally isolated from bacteria, have the unique ability to recognise specific sequences of bases within a DNA molecule and cleave the two DNA strands at a precise point within this base recognition sequence. This recognition and cleavage is illustrated in Figure 1, using as example the restriction endonuclease Eco RI. The restriction enzymes thus allow the large DNA molecules of higher eukaryotes to be broken down into much smaller but well-defined fragments, which are then amenable to experimental investigation.

Another important feature of the cleavage site of some restriction enzymes is also shown in Figure 1, and this is the formation of "sticky ends", which comes about in the following manner. The restriction enzyme recognition

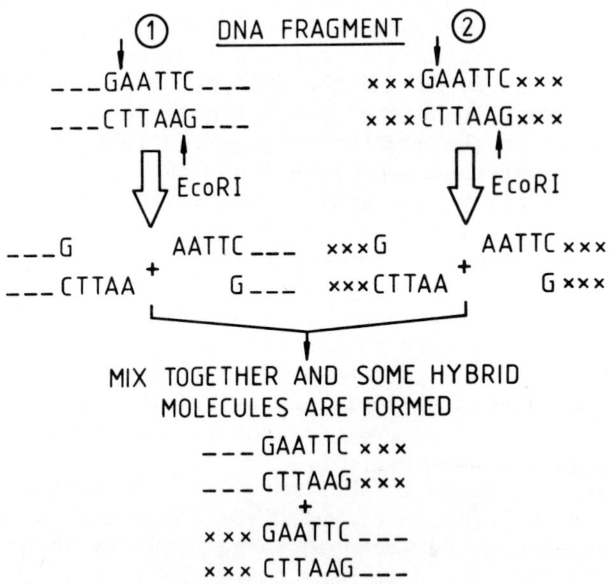

FIGURE 1. Diagrammatic representation of the cleavage of DNA by the restriction enzyme Eco RI.

sequence of one strand of a DNA molecule is palindromic with the complementary bases of the other strand. Thus, if the actual cleavage site is asymmetrically located within the recognition sequence, staggered ends will be formed at each end of the DNA fragment. These ends allow a precise method for the rejoining of two DNA molecules. Provided two DNA fragments have their "sticky ends" generated by the same restriction enzyme, when mixed together they will rejoin because of the complementary base-pairing properties of DNA. The two strands subsequently can be joined covalently by treatment with the enzyme DNA ligase. The important feature of this rejoining mechanism is that the procedure is independent of the initial source of DNA. Thus, fragments can be generated from two different sources of DNA using the same restriction enzyme, and when mixed, will give rise to hybrid molecules containing sequences from both DNA sources. This DNA is called "recombinant DNA" for obvious reasons, and is of vital importance for the new era of molecular genetics.

A DNA fragment which is required for experimental investigation can be produced in a highly-purified form in large quantity, by making a recombinant DNA molecule consisting of the DNA fragment and a bacterial plasmid. Plasmids are small pieces of bacterial DNA which replicate autonomously within bacteria, are not required for the survival of the bacteria, but which can carry certain properties such as antibiotic resistance genes which allow the bacteria to grow in the presence of those particular antibiotics. By using the procedures just outlined, it is possible to create a recombinant molecule which consists of a bacterial plasmid together with a piece of DNA which is required for investigation. This recombinant plasmid is then re-inserted into a bacterium and large quantities of the bacteria are grown in pure culture. The recombinant plasmid, which replicates along with the rest of bacterial DNA, can then be purified from the bacterial culture and the inserted piece of DNA recovered from the plasmid DNA by the same restriction enzyme used to insert it originally (Fig. 2).

Recombinant DNA molecules are also essential in the movement of isolated genes, by allowing the gene to be combined with a piece of DNA which facilitates its transfer from one organism to another. More will be said about this shortly, in discussion of the current situation with regard to the transfer of isolated genes between organisms.

The third technique which is involved in the revolution sweeping through molecular genetics is the ability to obtain the base sequence of DNA molecules simply and precisely. Two techniques have been developed for this purpose, one using

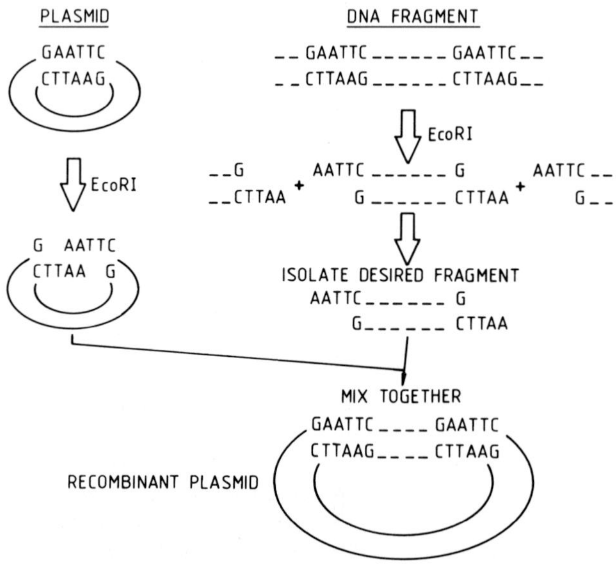

FIGURE 2. Diagrammatic representation of the preparation of a recombinant DNA molecule consisting of a bacterial plasmid and a piece of foreign DNA.

base-specific chemical degradation procedures (Maxam and Gilbert, 1980) and the other using an enzymic procedure in which nucleotide analogues are used in combination with a DNA polymerase (Sanger, Nicklen and Coulson, 1977). The short radioactive DNA fragments generated by either procedure are separated electrophoretically at very high resolution, and detected autoradiographically. Using these methods it is possible to obtain in relatively short time the base sequence of complete genes.

It is then apparent that molecular geneticists are now in a position to grow most eukaryotic cells in culture under well-defined conditions, to isolate DNA either from the organism under study or from cultured cells derived originally from this organism, to cleave this DNA at specific sites to produce experimentally manageable fragments, to recombine these fragments with any other desired piece of DNA, to obtain large amounts of any of the fragments as desired, and to obtain the complete base sequence of each DNA fragment of interest.

III The Structure of Eukaryotic Genes

The great potential engendered by these powerful new tools of research is fully demonstrated by examining the rapidly-increasing knowledge that has recently been gained about the structure of eukaryotic genes and their organization within the genome. Prior to the advent of the use of recombinant DNA, our concept of eukaryotic genes was largely based on the knowledge gained from prokaryotes. Thus a gene was considered to contain a structural element and one or more controlling elements, based on the operon concept of Jacob and Monod (1961). While it was well-recognised that the eukaryotic genes were arranged in a more complicated fashion than those of prokaryotes, hypothetical models for the control of the eukaryotic gene all used the operon concept as a base (Britten and Davidson, 1969; Davidson and Britten, 1979). It was therefore with some considerable surprise that it was found that the eukaryotic gene in fact is structured in a substantially different form to that of the prokaryotic gene.

The most striking difference in structure was the finding that many of the eukaryotic genes are "split" genes. By this is meant that the structural element, or coding region of the gene, is not a continuous sequence of nucleotides but is interrupted by long sequences of nucleotides that do not apparently code for polypeptides and whose function at present is unknown (Fig. 3). These interrupting sequences are known as "introns", or "intervening sequences", and can vary in number from one (actin genes) to at least thirty-three (vitellogenin genes). Upon expression of these genes, RNA transcripts of the entire gene, including intervening sequences are made, and the intervening sequences are subsequently removed by a "splicing enzyme", thereby giving rise to a mRNA with a continuous coding region ready for translation into a polypeptide. The mechanism whereby the splicing enzyme recognises the intervening sequences is not yet clear, although the junctions between the intervening sequences and the coding sequences do contain consensus sequences (Breathnach and Chambon, 1981) which is thought to be an important part of the recognition process. However,

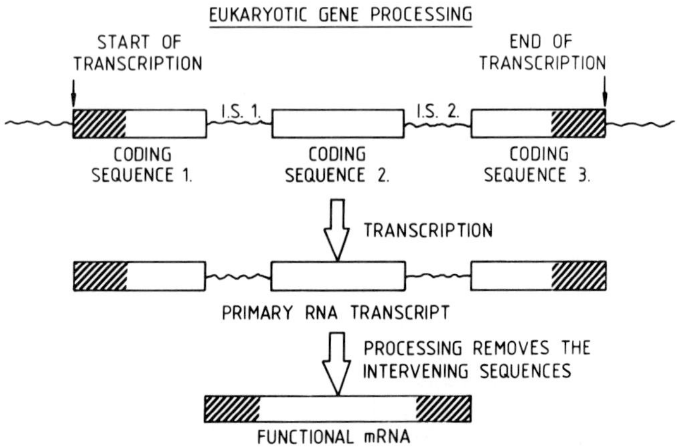

FIGURE 3. The hypothetical structure of a eukaryotic gene. (IS) = intervening sequence.

not all eukaryotic genes contain intervening sequences. For example, the coding sequences of the histone genes are not interrupted, and more recently it would appear that the coding region of at least some of the wool keratin genes (Powell *et al.*, in preparation) and avian keratin genes (Molloy *et al.*, in preparation) do not contain intervening sequences.

In line with the basic concept of the bacterial operon, the grouping of some coding sequences which are functionally associated has been observed. For example, the grouping of the various hemoglobin genes has been clearly demonstrated for a number of species, and the order in which the genes are arranged appears to reflect the order of their expression during development. It is also of interest to note that the functional hemoglobin genes are interspersed with genes which apparently are not functional, but which bear great similarities to adjacent functional hemoglobin genes. These are called "pseudo genes" and are possibly some mutational variant of the functional gene arrived at during the course of evolution. It will be of interest to see to what extent the wool keratin genes, whose proteins have much less structural constraint on them than most other proteins, show "pseudo gene" characteristics. It may be that the "pseudo genes" are still expressed as minor variants of the original gene, thereby giving rise to the complex multi-gene family structure which appears to be in operation from the protein-structural studies which have been carried out on the wool keratins.

While this review of our current knowledge of the structure of the eukaryotic gene is necessarily brief, it should be apparent that our knowledge has increased greatly in the last few years and the rate of acquisition of this knowledge shows no real sign at present of slowing down. It does not seem unreasonable therefore, to expect that within the next few years the basic structure of the eukaryote gene, including its controlling elements and processing features, will be largely understood. From the view point of this symposium, we may then ask, how can this mass of new knowledge best be applied to the field of animal production?

It would seem that two major problems bar the way to progress in the improvement of animal production characteristics by conventional methods of genetic selection. The first of these is the long generation time of our domestic animals. This inevitably results in experimental breeding programmes which are extended over a large number of years, because of the long gestation period of the animals and then the long period required before this animal can produce off-spring. The drawbacks of long breeding programmes are the difficulty in maintaining relatively standard environmental conditions for long periods of time, and of maintaining experimental breeding herds over long periods. In addition, such programmes rapidly become very expensive because of the large number of animals which are required for adequate statistical analyses of the experimental results.

The second main problem barring the way to animal production improvement is the difficulty of moving a desired production characteristic from one strain of animal to another without at the same time moving one or more undesirable characteristics. The reasons for this are probably many in number. Most production characteristics are not the result of single gene action, but are the phenotypic expression of a number of genes acting in concert. Thus, the transposition of such a characteristic involves the movement of all the relevant genes. Because of the mechanism of gene transfer in higher organisms, which involves the movement of whole chromosomes or at least of large portions of chromosomes, hundreds of genes are inevitably transferred along with the desired gene. Since production characteristics can involve many genes, their transfer must involve the simultaneous transfer of a huge amount of genetic material not relevant to the production characteristics. Should any of this material be associated with an undesired phenotypic property, the transfer of this property will also occur. The closer the physical association of the undesired genetic information with the relevant genes, the more difficult it is to avoid the simultaneous transfer of both characteristics.

IV Techniques for Gene Transfer

The new techniques available in molecular genetics would seem to provide a possible way of circumventing these drawbacks to conventional breeding procedures, by transferring, from one animal to another, single genes or groups of desired genes which are isolated free of genes which give rise to undesirable characteristics. This approach is obviously dependent on the availability of methods for the actual transfer of isolated genes from one eukaryotic cell to another. Fortunately there are now three different methods which have proven effective for such transfer operations. These are the use of vectors derived from transforming viruses, the transformation of cell culture lines with naked DNA and the direct microinjection of purified DNA.

The use of vectors derived from transforming viruses has proved very effective in facilitating the transfer of specific genes. For example, a number of vectors have been constructed from the SV40 viral genome which have incorporated the bacterial gene for the enzyme xanthine-guanine phosphoribosyltransferase (Mulligan and Berg, 1980). These vectors allow the introduction of specific genes into cell culture lines and the selection of the transformed cells by growth on suitable selection media.

The transformation of tissue culture cells by naked purified DNA has been demonstrated (for review see Pellicer *et al.*, 1980). Particularly useful has been the technique of co-transformation whereby a DNA sequence containing a selectable marker is mixed with a nonselectable gene and used to transform a cell culture line. Most cells recovered by reason of their transformation with the selectable marker are also found to contain the DNA of the nonselectable gene as well.

The two procedures described above are successful in introducing foreign DNA into eukaryotic cells and the DNA is stably and covalently incorporated into the recipient genome.

Furthermore, in many instances the genes of the inserted DNA appear to be expressed in the new cell environment. However, it has not yet been possible with either technique to demonstrate that the expression of the introduced genes is under appropriate control by the cell.

The microinjection of naked DNA has recently proved to be the most successful of the techniques for the transfer of isolated DNA. Initial experiments using fertilized frogs' eggs (Bendig, 1981) demonstrated that such injected DNA was stably incorporated into the cell and was expressed at high efficiency. More recently, the technique received a major impetus with the demonstration that naked DNA, injected into mouse pronuclei, was stably incorporated and more importantly was carried through into the progeny (Constantini and Lacy, 1981; Wagner *et al.*, 1981; Brinster *et al.*, 1981). The rabbit β-globin genes inserted by Constantini and Lacy (1981) were not expressed in the tissues of the recipient mice. However, the other two research groups have both demonstrated that the inserted genes could produce functional proteins. Thus Wagner *et al.* (1981), using the rabbit β-globin gene, reported the production of rabbit β-globin in their progeny mice, although an apparent overproduction of the protein by these animals suggested that the mouse might not be able to properly control the foreign rabbit genes. On the other hand, Brinster *et al.* (1981) in their study used a gene constructed by fusion of a mouse metallothionein gene regulatory sequence and a herpes simplex virus structural gene, and demonstrated firstly that the foreign gene was expressed in the mice, secondly, that this expression was subject to regulation by heavy metals and thirdly, that it was expressed in liver and kidney tissues but not in brain, which parallels the expression of the metallothionein genes themselves in these tissues. These reports suggest that a major obstacle to the use of the new technology, the insertion of naked DNA into a recipient organism in such a position that it is stably inherited and under correct cellular control, is partially overcome. Since techniques exist for the culture, growth and manipulation of oocytes and fertilized eggs from most, if not all, of our major domestic species of animals, the transfer of isolated genes from one breed to another would in principle now appear to be experimentally feasible.

V Gene Manipulation and Animal Production

The preceeding summary of our current position with regard to the techniques for gene isolation and transfer leads us to ask whether we can identify pieces of DNA which have a role in animal production, and which, when transferred to another strain, alter the recipient animal in the desired fashion. The genetic components with which we must concern ourselves can be broadly divided into two groups: characteristics which are controlled by single genes, and characteristics which are controlled by a complement of genes acting in concert. Of these two categories, it is obvious that the first would be much easier to handle with our current knowledge of eukaryotic gene structure.

We can envisage the control of animal production characteristics by single genes operating either at the structural level, where differences in the actual protein component of the animal are altered, or at the level of gene control, where the expressed levels of specific protein components are altered. The alteration of the structural elements of a specific gene is the easiest of these targets. What then can we expect to achieve in this area? The major structural proteins important in the area of domestic animal production are the textile fabrics (e.g., wool, cashmere, mohair), the proteins which give rise to muscle fibres (e.g., actin, myosin) and the proteins which are involved in milk production (e.g., caseins, lactoglobulins).

As an example, let us take the wool fibre proteins for more detailed examination. The wool fibre is made of a complex mixture of proteins called keratins, which can be arranged in two major groups according to their amino acid compositions (Crewther, 1976). Within these groups, the individual proteins fall into families of polypeptides which have highly-conserved amino acid sequences within each family. The proteins are tightly bonded together in the dead cells of wool fibres and form a compact structure of microfibrils and surrounding matrix. The protein properties suggest that the genes which encode the keratins may be arranged as multi-gene families, which are defined as groups of genes which are physically grouped and which encode polypeptides of similar amino acid sequences. Recently, studies have been initiated to determine whether this postulated arrangement of the wool keratin genes is indeed correct.

In order to establish molecular probes for the genes, wool keratin mRNA was obtained (Ward and Kasmarik, 1980) and a comprehensive library of keratin complementary DNA (cDNA)

TABLE I. Identification of six Keratin cDNA Clones with known Wool Keratin Proteins

Clone	*Size (bp)*	*Keratin protein encoded*	*Fraction of coding sequence (%)*
09	*800*	*SCMKA-7C (MICROFIBRIL)*	*(30)*
V15	*850*	*SCMKA-8C$_1$ (MICROFIBRIL)*	*(60)*
SK10	*500*	*SCMKB-2C (MATRIX)*	*23*
D31	*500*	*SCMKB-IIIB4 (MATRIX)*	*15*
KS18	*600*	*SCMKB-IIIB3 (MATRIX)*	*100*
KS101	*500*	*SCMKB-IIIB2 (MATRIX)*	*100*

A population of double-strand cDNA molecules copied from wool keratin mRNA was cloned in the Pstl site of the plasmid pBR322 and used to transform the host E. coli K12 strain RR1. Clones which were positive to hybridization with ^{32}P-keratin cDNA were sequenced by the method of Maxam and Gilbert (1980) and the amino acid sequences generated were compared with known wool keratin amino acid sequences.

() in the Fraction of coding sequence values indicate some uncertainty still exists due to a lack of protein sequence data.

clones has been prepared and characterized (Table I) (Sleigh and Ward, in preparation). These probes have been used to screen for their corresponding gene families in a sheep cloned genomic DNA library. So far, only one of the many clones identified has been examined in detail (Powell, Sleigh and Ward, in preparation). It has been shown to contain two keratin genes (Fig. 4) located adjacent to one another, with about 2 kilobase pairs separating them. As predicted from the expected multi-gene arrangement of the keratin genes, these two genes both belong to the same keratin family, the high sulphur II group. It does not seem unreasonable therefore to expect that the keratin genes will be grouped and that they will be controlled by regulatory sequences which will probably (although not necessarily) be located close to the structural gene batteries.

The amount of wool grown in a defined time is primarily dependent on the rate of keratin protein synthesis. However, to obtain more wool, it is not necessary to uniformly increase the rate of synthesis of the entire keratin protein spectrum. For example, the keratins in general are charact-

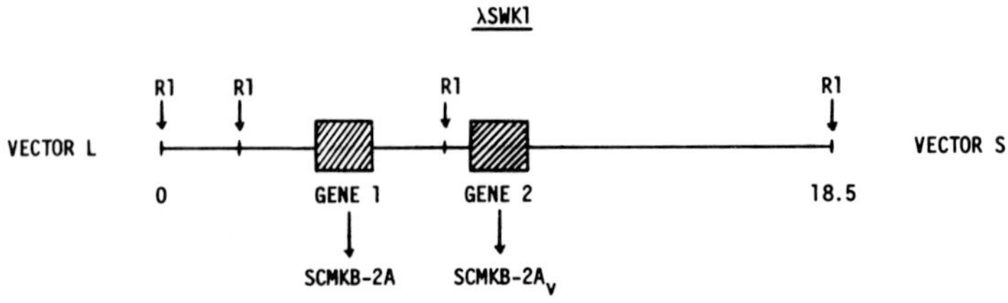

FIGURE 4. The structure of a wool keratin genomic clone λ SWKI containing two adjacent keratin genes coding for members of the high sulphur II keratin family.
(RI) = Eco RI restriction site.
(VECTOR L, VECTOR S) = long and short arms respectively of the λ Charon 4A phage vector. Numbers represent kilobase pairs.

erised by a high content of the amino acid cysteine, and when the level of cysteine circulating in the blood of a sheep is experimentally increased, the rate of wool growth is also increased (Reis and Schinkel, 1963; Reis *et al.*, 1973a, b). When the composition of the wool fibres which have been grown during the period of increased rate of synthesis is examined, it is found that the high-sulphur keratins have been increased greatly with respect to the low-sulphur keratins. Furthermore, a group of proteins called the "ultra-high sulphur" keratins are synthesized in large amounts whereas normally these proteins are a very minor component of wool. It follows therefore that a method which allows for an increase in the levels of these proteins would result in an increase in wool growth. It is entirely possible that by increasing the level of the gene copy number for these proteins in the sheep genome, one may be able to induce the desired increase in protein synthesis, given a suitable nutritional regime for the animal capable of supporting the additional protein synthetic requirements. Alternatively, it may be advantageous to reduce the level of cysteine present in the major keratin proteins, provided that the reduction did not proceed to the point where the physical properties of the fibre itself were impaired. The effect of reducing the requirement of cysteine would presumably allow an increased rate of protein synthesis under conditions where cysteine is limiting.

Changes in the amino acid composition (e.g., cysteine content) may also benefit the industry by altering the properties of the wool fibre in a beneficial fashion. Thus for example it may be possible to change the elasticity of fibres by altering the extent of crosslinking which is present due to the disulphide bonding of the cysteine residues of the keratins.

These examples show the potential which exists for product and productivity gains in wool biosynthesis by directly altering the structural component of the wool genes. However, greater productivity changes are potentially achievable by alterations made to the regulatory component of the wool keratin complex. In this area, however, we have moved one step further into the realm of future speculation, because the regulatory sequences of genes are not yet understood. Nevertheless, it is clear that some regions of all eukaryotic genes are involved in controlling expression. The region about 20 to 30 bases upstream from the initiation of transcription contains an AT-rich base sequence region known as the TATA box (Gannon *et al.*, 1979), which is involved in the mechanism of transcription. The role of introns has not yet been clearly defined, and these may also play an important role in governing the amount of mRNA which is produced from a particular gene at any time. When the regulatory mechanisms concerned with transcriptional control are better understood, it should be feasible to transfer regulatory sequences which favour high levels of keratin production to animals which have desirable wool fibre properties but low levels of production. Alternatively, it may also be feasible to isolate the regulatory sequences, modify them by changing appropriately selected bases and then re-insert them into the genome from which they were isolated, or alternatively into a new genome, in order to allow altered rates of keratin synthesis. As was the case with the situation when structural modifications were made to the genes, suitable nutritional modifications would have to be made to cope with increased keratin synthesis. Also, it is possible that by changing the rate of keratin protein synthesis, changes in the ratios of the various component keratins may occur, which could lead to changes in the physical properties of the resultant wool fibres.

Another role for modification directed at the controlling elements of the keratin genes would be the addition of a functional regulatory sequence which responds to a repressor molecule not normally encountered by the animal. In this way, keratin gene transcription would be amenable to direct manipulation by administration of the appropriate repressor substance. This would be of great value to the wool industry in the area of wool harvesting, since chemical defleecing has for some time been considered to be a most attractive option as a method of wool removal, but is not practical at present because of the unavailability of suitable chemicals with substantial specificity for the wool follicle.

The potential for wool fibre production improvement by the new procedures of molecular genetics also obviously exists for the other major areas in animal production, i.e., muscle growth and milk production. In the case of muscle growth, similar arguments exist for the proteins actin and myosin as for the keratin proteins. However, it is unlikely that significant structural changes can be made in the muscle proteins because they have such a vital functional role to play in the body, and it is reasonable to assume that they have become very well adapted for this role. The wool keratins would be expected to be under much less constraint because their function is restricted to the ability to form an adequate cross-linked insoluble protein complex. Thus more importance should be attached to the possibilities of altering the regulatory sequences which control the production of actin and myosin.

Milk production is a complex series of events which involve not only protein biosynthesis but also carbohydrate and lipid synthesis as well. More will be said shortly about this when we consider those characteristics controlled by multiple genes. Nevertheless, while milk as a whole is clearly the result of the expression of many genes, it is reasonable to assume that increasing the production of a single component of the secretion may well raise the level of production of the entire liquid. Thus increases in one or more of the casein proteins may be advantageous, and the principles discussed for the wool keratins in general would apply to these milk proteins. Some of the other important milk proteins also have an enzymic role as well as a passive nutritional role, and as in the actin/myosin situation the genes for these proteins would be likely to be candidates more for regulatory alteration than for structural changes. Similarly, the large number of enzymes which are involved in carbohydrate and lipid syntheses would be likely to be amenable to alteration of their levels rather than to structural changes in their actual amino acid sequence.

The production characteristics which are controlled by single genes are in the minority, most characteristics being the phenotypic expression of the concerted action of a number of genes. This applies to some extent to some of the factors involved in muscle growth and milk production, but much more to important factors in animal production such as reproductive performance and nutritional performance.

While some of the control of muscle growth can be considered under single gene control, as discussed earlier, it is apparent that the totality of muscle fibre formation is a complex interaction of biochemical events. The formation of muscle cells is initially defined during development and is dependent on rates of cell division and the as yet unknown tissue-specific factors which result in the selection and differentiation of a particular group of cells. Once the myogenic lines have been established, cell growth is dependent on the numerous enzyme activities involved in DNA replication, RNA synthesis, energy supply and protein synthesis. Finally, as terminal differentiation of the muscle cell is approached, synthesis of the muscle-specific proteins becomes a critical factor in determining the size of the muscle cell. However, even at this stage, the numerous enzymes involved in the formation of cell membranes and outer cell envelopes must be functioning actively.

Similarly, milk production can best be considered as being under the control of numerous enzyme systems and hence under the influence of multiple genes. Thus the protein, lipid and carbohydrate components no doubt all influence the rate of milk production, as discussed earlier, and it is apparent that there exists a complex regulatory system to control these components, because each of them can be varied to a greater or lesser extent quite independently of the others during the secretion of milk at different stages of lactation. Milk production is also extensively controlled by hormonal levels (e.g., prolactin, steroids) and the genes which control these hormones, as well as the genes which control the levels of the hormone receptors in the mammary tissue cells, would therefore be expected to be involved in the control of milk production. In addition, the actual amount of mammary gland tissue which is present and available for milk production is of considerable importance, and this involves the large number of gene sets which are required during the process of differentiation and development of mammary gland.

If the genetic control of muscle fibres or milk production gives the appearance of being complex, such complexity pales into insignificance when considered next to the control of functions such as reproductive and nutritional performance. Let us first consider the control of reproductive performance. While this remains only partly understood, it is clear that reproduction is controlled by the interaction of the products of a very large number of genes. The hormonal interactions alone involve three groups of hormones, the peptide releasing hormones, the pituitary protein hormones and the steroid hormones. For each relevant hormone there is at least one gene involved in its synthesis, and in many cases a number of genes are involved. In addition, each hormone has associated with it a tissue-specific receptor protein which again presumably requires at least one gene for each receptor. Numerous different tissues are involved in the process of reproduction. The urogenital tract of both male and female animals undoubtedly requires the concerted, co-ordinated action of a large number of genes. During pregnancy, not only are the genes of the mother required for the maintenance and growth of the foetus, but the foetal genes themselves become operative in pregnancy maintenance as well (e.g., placental growth). Finally, the developmental programme which is followed by the growing foetus is specified by the foetal gene complement.

The genetic control of nutrition also presents a picture of bewildering complexity to the molecular geneticist. Ingested food must be degraded both chemically and enzymically by the concerted action of many enzymes, each of which involves at least one gene. The products released by digestion must then be transferred into the blood stream of the animal, which involves a complex transport system across the gut lining. The secretion of digestive enzymes and the amount of food ingested by an animal are both under poorly understood hormonal controls, which presumably involve additional genes. In addition, many of our important domestic breeds of animals are ruminants, and hence genes involved in the formation and maintenance of the rumen must also be considered in the context of the genetic control of nutrition.

It is apparent from the preceeding survey of the role of multiple gene systems in animal production, that the complexities of these systems place them outside the realm of molecular genetics at its current level of knowledge. Thus there is very little prospect of direct manipulation of these functions by the new techniques of molecular genetics at the present time. Indeed, it is unlikely whether many of the genes involved could even be identified with our current knowledge. However, it is equally apparent that the major areas where improved animal production performance might be expedited are intimately involved with these complex genetic batteries. How then can we envisage the molecular geneticist playing a role in this area? Initially, two advances in knowledge must be made. Firstly, the genetic complexity of reproductive performance must be divided into manageable segments, by isolating some segments that might be more important than others in influencing the overall function. For example, the gene bank involved in the production of the hypothalamic releasing hormones might prove to have such a critical role in the whole reproductive process that small changes in one or a few of the constituent genes of this gene bank may elicit major changes in the whole process. Areas where these major changes might be useful are in the increase of ovulation rate and hence number of embryos, or by aiding pregnancy maintenance and hence decreasing the incidence of embryo mortality. Secondly, methods must be available for the actual physical identification within a gene bank of the gene which is of importance, so that the appropriate piece of DNA can be physically isolated. Where the gene is responsible for a recognisable product such as an enzyme or a hormone, this is currently no more than a technical problem whose difficulty is determined by the level of expression of the gene. However, it is quite possible that some genes do not give rise to recognisable or isolatable gene products, but instead may be entirely regulatory in nature. It would be very difficult with our current knowledge to isolate such genes unless they were closely associated with genes which could be clearly identified and isolated. Even then, their recognition would be difficult at present. Nevertheless, assuming that within a gene subset a gene could be identified which had a significant influence on the overall performance of a function such as reproduction, this gene would be transferrable by *in vitro* means essentially as discussed earlier for the transfer of single keratin genes. As our understanding of the interplay of entire gene subsets becomes better understood, it would then be possible to consider the transfer of such entire subsets, suitably modified *in vitro* if desired before transfer. It would then be possible to move numbers of relevant genes from one genome to another without the complication mentioned earlier of the simultaneous transfer of a huge number of extraneous and often unwanted genes. Nevertheless, it would be expected that as the number of transferred genes was increased, the incidence of unexpected and sometimes undesired phenotypic expressions might increase because of unexpected interactions between the new genes and the recipient genome. These would be expected to be much less, however, than those which occur using conventional breeding techniques.

VI Gene Manipulation and the Rumen

A role for molecular genetics in the improvement of animal production which is not part of the animal genetic system itself exists because of the ruminant nature of many of our domestically useful breeds. The rumen is best considered as a large fermentation vessel which contains a complex bacterial, protozoal and fungal population growing in largely steady-state conditions. When food is ingested by ruminants it is degraded to a large degree in the rumen and many of the component nutrients which are made available are utilized for growth by the rumen microflora. Because of the steady-state growth conditions of the rumen, the continuously produced supply of micro-organisms is constantly flowing from the rumen to the more distal regions of the animal's digestive tract where it is degraded by digestive enzymes and the nutrients absorbed by conventional procedures. In addition, fermentation products such as volatile fatty acids are produced in the rumen from the metabolic processes of the microflora, and these are also utilized by the animal as a nutritive source.

It is clear from the above description that the ruminant animal is dependent for many of its nutrients entirely on the extent to which they are produced by the rumen microflora. Hence, if a particular nutrient is known to be a rate-limiting factor in a particular biosynthetic process of the animal, methods which increase the supply of this factor in the rumen should result in an increase in the associated biosynthetic process. It is here that molecular genetics has a clear role. The bacterial population of the rumen is well-adapted to its environment, and it is very difficult to introduce other bacteria into the population in a way which allows them to form a stable colonization. However, there are numerous genes in non-ruminant bacteria that would be of great advantage to ruminant bacteria in terms of their provision of nutrients to the animal itself. For example, bacterial genes which increase the availability of the amino acid cysteine to a sheep, either by synthesis and secretion of the amino acid or by incorporation of the amino acid into high-cysteine bacterial proteins, would be of great value to a sheep in increasing wool growth.

It is not conceptually difficult to envisage the transfer of a selected gene of obvious ruminant advantage from the genome of a non-ruminant bacterium to the genome of a ruminant bacterium, and the re-introduction of the modified bacterium back to the rumen. Clearly it would be hoped that the introduction of the new gene to the bacterium did not confer a disadvantage on the survival and colonization properties of the bacterium in the rumen. However, even if this did occur, microbiological techniques now exist which might allow such a problem to be overcome. If, after genetic modification, the bacterium could not form a stable colonization of the rumen, another approach to its utilization would be to grow pure cultures of the bacterium and inoculate the sheep rumen with the bacteria at suitable intervals. It would be predicted that the concentration of some of the nutritional components which are rate-limiting to animal production might be increased significantly by this method.

VII Gene Detection and Animal Breeding

An area where the new techniques of molecular genetics could be of value in conventional animal breeding procedures is in the detection of genes in the genome of an animal without first having to wait for the animal to mate and produce offspring. This relies on the gene in question being known either by its DNA sequence or by the availability of a nucleic acid probe which can recognize the gene. A small sample of tissue from the animal in question could then be collected, DNA prepared and after suitable restriction enzyme digestion, separated by electrophoresis. Hybridization of the separated DNA with an appropriate nucleic acid probe would then detect the presence of the particular gene in question. Of course, the presence of the gene in the genome does not automatically ensure its expression in the phenotype of the animal, but some of this information may conceivably be available from detailed knowledge of the gene's behaviour obtained by conventional genetic crossing data. The extension of this procedure to the detection of sets of genes would be relatively simple, but interpretation of the known genetic composition in terms of phenotype would rapidly become very difficult. However, the procedure appears to offer an approach to the screening of many animals for the presence of a particular gene sequence.

VIII Conclusion

In conclusion, it would seem that the new techniques available in the field of molecular genetics are going to play a revolutionary role in animal breeding and selection in the future. We now have available to us procedures for the identification and isolation of individual genes, the transfer of these genes between the genomes of animals and the growth of new animals containing the modified genomes. As our knowledge increases in the area of gene function and control, it should be possible to isolate and transfer co-ordinated sets of genes which control or influence such complicated physiological processes as reproduction or nutrition. It will be possible to manipulate the genetics of the rumen bacterial microflora as desired and to detect the presence of specific genes or gene sets in herds of domestic animals. Finally, it will be possible to grow large numbers of genetically identical animals of a desired phenotype by isolation of desired nuclei and their introduction into enucleated oocytes. When this is achieved, animal production techniques will bear very little resemblance to those currently in operation today.

REFERENCES

Bendig, M.M. (1981). *Nature, Lond. 292,* 65.

Breathnach, R., and Chambon, P. (1981). *A. Rev. Biochem. 50,* 349.

Brinster, R.L., Chen, H.Y., Trumbauer, M., Senear, A.W., Warren, R., and Palmiter, R.D. (1981). *Cell 27,* 223.

Britten, R.J., and Davidson, E.H. (1969). *Science 165,* 349.

Constantini, F., and Lacy, E. (1981). *Nature, Lond. 294,* 92.

Crewther, W.G. (1976). *In* "Proceedings of 5th International Wool Textile Conference", Aachen *1,* 1.

Davidson, E.H., and Britten, R.J. (1979). *Science 204,* 1052.

Gannon, F., O'Hare, K., Perrin, F., LePennec, J.P., Benoist, C., Cochet, M., Breathnach, R., Royal, A., Garapin, A., Cami, B., and Chambon, P. (1979). *Nature, Lond. 278,* 428.

Jacob, F., and Monod, J. (1961). *J. molec. Biol. 3,* 318.

Maxam, A.M., and Gilbert, W. (1980). *In* "Methods in Enzymology" (L. Grossman and K. Moldave, eds.), p. 499. Academic Press, New York.

Mulligan, R.C., and Berg, P. (1980). *Science 209,* 1422.

Pellicer, A., Robins, D., Wold, B., Sweet, R., Jackson, J., Lowy, I., Roberts, J.M., Sim, G.K., Silverstein, S., and Axel, R. (1980). *Science 209,* 1414.

Reis, P.J., and Schinkel, P.G. (1963). *Aust. J. biol. Sci. 16,* 218.

Reis, P.J., Tunks, D.A., and Downes, A.M. (1973a). *Aust. J. biol. Sci. 26*, 249.
Reis, P.J., Tunks, D.A., and Sharry, L.F. (1973b). *Aust. J. biol. Sci. 26*, 635.
Sanger, F., Nicklen, S., and Coulson, R. (1977). *Proc. natn. Acad. Sci. U.S.A. 74*, 5463.
Wagner, T.E., Hoppe, P.C., Jollick, J.D., Scholl, D.R., Hodinka, R.L., and Gault, J.B. (1981). *Proc. natn. Acad. Sci. U.S.A. 78*, 6376.
Ward, K.A., and Kasmarik, S.E. (1980). *J. invest. Derm. 75*, 244.

Discussion Summary

The prospect of being able to remove, alter and replace genes was seen to be exciting, while the opportunity of synthesizing genes directly and then inserting them would open up many possibilities. A model was proposed whereby 'mini chromosomes' be produced by extracting DNA fragments from animal cells, transferring into an organism or cell with a high turnover rate, and then selecting and purifying specific gene subsets. Having produced the desired combinations of subsets, the 'mini chromosome' would be returned to the donor species. There were doubts as to the feasibility of this procedure because of the large number of initial DNA fragments, including those containing undesirable genetic material, and because of the random rejoining of the fragments in culture.

The known variation in intervening sequences within a species is very small. This may be an indication that they have been maintained during evolution for some reason. It is known that these intervening sequences play a functional role during transcription. However, there is considerable variation between species, genera, phyla, etc. in the structural genes as well as the intervening sequences. None of this variation has as yet been ascribed to phenotypic differences. It will have implications for the transfer of DNA across species, etc.

Genetic modification of rumen microflora was seen as an attainable goal in the short-term and as being potentially important to commercial production, particularly where the ruminant was not perfectly adapted to the feed sources.

The major point emerging from the discussion was that there is an acute shortage of information on the physiological control of the productive processes in animals, e.g., milk production, growth. By the time genes and gene products are identified for more of the regulatory steps in these processes, the molecular biologists should be able to isolate these genes and manipulate them to the future advantage of the animal breeder. Our very limited knowledge of the regulation of structural genes in eukaryotes was seen as an important restriction to progress.

Part 3
Immunogenetics

The Future Role of Immunogenetics in Animal Breeding

M. J. Stear

Department of Immunology
The John Curtin School of Medical Research
Australian National University
Canberra

This paper speculates on the influence that immunogenetics will have on animal breeding in the future and predicts that breeding for disease resistance will become easier in the next few years. We will be able to identify genes which control disease resistance and to use selection programmes to optimize gene frequencies. Additionally, an improved ability to detect individual genes will allow better detection of parentage errors, more accurate determination of breed relationships and better detection of deleterious genes in the heterozygous state.

I Breeding for Disease Resistance

Genetic variation in disease resistance has been shown for a large number of diseases in a variety of different species including animals of economic importance (reviewed by Hutt, 1958; Spooner *et al.*, 1975). However, for animal breeders the important point is not whether genetic variation in disease resistance exists but whether it is possible to improve disease resistance by selective breeding. For example, if superior disease resistance is due to overdominance, then breeding from the resistant animals (i.e., the heterozygotes) will not improve the average disease

FUTURE DEVELOPMENTS IN THE
GENETIC IMPROVEMENT OF ANIMALS
ISBN 0 12 078830 6

resistance of the population. Studies on laboratory rodents have shown that it is possible to improve resistance to specific diseases by selection (e.g., Biozzi *et al.*, 1975). Unfortunately studies on laboratory animals do not allow us to predict results in economically important livestock, as it is likely that the disease resistance of laboratory populations differs markedly from that of the wild populations from which the laboratory animals were derived. Selective breeding may merely restore the disease resistance level to that of the original wild population.

The results obtained with livestock are not extensive but they are encouraging. Hutt reviewed the field in 1958. In pigs, selective breeding improved resistance to brucellosis (Cameron *et al.*, 1942) and to erysipelas (Fortner, 1949). In chickens, breeding improved resistance to pullorum disease (Roberts and Card, 1935), fowl typhoid (Lambert, 1932), avian leucosis complex (Hutt, 1958) and coccidiosis (Rosenberg *et al.*, 1954). In sheep, breeding from resistant stock improved resistance to *Haemonchus contortus* (Warwick *et al.*, 1949), infectious pulmonary adenomatosis (Dungal *et al.*, 1938) and *Trichostrongylus colubriformis* (Windon and Duneen, 1981). Breeding from dams resistant to bovine mastitis produced a smaller proportion of susceptible offspring than breeding from susceptible dams (Ward and Castle, 1945; Legates and Grinnels, 1952). Resistance to the cattle tick *Boophilus microplus* can also be improved by selective breeding (Utech *et al.*, 1978).

The results are in many ways surprising. The expectation is that disease resistance would be maintained at an optimum by natural selection and disease resistance could not therefore be improved by selective breeding. However, the facts show otherwise. There are several possible explanations. It may be that disease resistance in many populations studied is not at equilibrium. The relationship between the immune system and potential pathogens is a continually evolving one. Artificial selection for production characters may slow down the rate at which natural selection produces the equilibrium, although if healthy animals are better producers, artificial selection may increase the rate of progress to equilibrium. Also, if natural selection against disease susceptible animals has been relaxed for any reason, this could prevent a population achieving its maximum disease resistance. Causes of a relaxation in natural selection would include dipping, vaccination and antibiotic therapy. Animals transferred to a new environment, including stock imported from abroad or transferred from a different climate, would need time to allow the population to achieve its optimal disease resistance. Even after a population has achieved equilibrium, deleterious mutations will continue to occur and occasionally, by chance, accumulate. Artificial selection may have a role to play in helping to eliminate the carriers of deleterious genes. Additionally, a distinction must be made between an endemic disease where the disease is regularly present in a community and an epidemic disease which only occurs in outbreaks. If an epidemic disease only occurs at intervals of many generations, natural selection will be unable to maintain resistance to this disease at the optimum. Finally, if different environments exist within a species range then any gene flow by transfer of embryos, semen or breeding stock will prevent adaptation occurring and prevent disease resistance achieving its optimal level.

It may also be possible to improve resistance to a specific disease in a population at equilibrium. A strong immune response prevents disease in many circumstances but in other circumstances a less vigorous immune response would be beneficial. Oldstone *et al.* (1973) have shown that lymphocyte choriomeningitis in mice results from an excessive immune response to a virus. Many autoimmune diseases are also thought to be due to overactive immune responses. It may be possible to increase the resistance to an economically important disease at the cost of increased susceptibility to other less important diseases.

However, there are difficulties in breeding for disease resistance. It is very inefficient to breed for disease resistance directly as this means giving all breeding stock the very disease that you are trying to prevent. One possibility which has been discussed by Spooner *et al.* (1975) is to select for longevity. This method has the advantage that part of the reason for improving disease resistance is to increase the useful life of livestock. The assumption is that longer-lived animals have, on the average, an increased disease resistance. This is probably true both with natural populations and with domestic stock which are culled on the breeder's assessment of their economic value. For example, Robertson and Barker (1966) have shown that longevity in dairy cattle has a genetic component which is independent of milk yield, although milk yield is by far the most important genetic determinant of survival. However, longevity is difficult to estimate at an early age.

The approach which was favoured by Spooner *et al.* (1975) was to identify specific genes which controlled disease resistance and to improve disease resistance by changing the frequencies of these genes. The approach has been made feasible by recent advances in immunogenetics. Given the wide variety of host defence mechanisms and the genetic complexity of potential pathogens, it seems unlikely that genes controlling host resistance could be detected. Nonetheless, genes involved in host resistance have been identified in humans (Dausset and Svejgaard, 1977), mice (Lilley, 1966), rats (Gasser *et al.*, 1973; Williams and Moore, 1973), chickens (Longenecker and Mosman, 1981), pigs (Sellwood, 1979), sheep (Windon and Duneen, 1981) and cattle (Stear *et al.*, unpublished). Nor have genes only been identified in specific types of disease. Genes have been identified which control resistance to autoimmune diseases (Ryder *et al.*, 1979) and to diseases caused by viruses, bacteria, protozoa and metazoan parasites (Skamene *et al.*, 1980). These results suggest that the control of disease resistance is less complex than originally believed, which may have implications for other multifactorial traits.

The most interesting of the genetic systems which control disease resistance is the major histocompatibility system (MHS). This is a genetic system, consisting of several genes in close linkage, which regulates the immune system (Benacerraf, 1981). It is present in all species of higher vertebrates so far studied (Götze, 1977). Diseases in which the immune system plays a role, including cancers, infectious diseases, autoimmune diseases and allergies, predominate in the list of known MHS associated diseases. Interestingly, many diseases of unknown aetiology which have been shown to be associated with the MHS are now assumed to be immunologically mediated. The associations observed with the MHS are more exciting than those observed with red blood cell systems for two reasons. Firstly, there is a rational explanation for the associations. It is reasonable that genes which regulate the immune response should be associated with disease resistance. Secondly, the associations observed between the MHS and specific diseases are very much higher than those observed for red blood cell systems and disease. If the strong associations observed in humans also occur in livestock, then it will be feasible to improve disease resistance using only the MHS as a disease resistance marker.

Table I shows the relative risks for several selected diseases in humans. For example, humans with the two alleles DR3 and DR4 have a much greater chance of contracting insulin dependent diabetes mellitus than individuals who lack these two antigens. The relative risk measures how many more times a disease is likely to occur in individuals possessing a certain antigen than in individuals not possessing that same antigen. It is calculated as:

TABLE I. Relative Risks for Selected Diseases with Genes of the Human Major Histocompatibility System

Disease	*Antigen*	*Relative Risk*
Insulin-dependent Diabetes Mellitus	*DR 3,4*	*33.1*[a]
Multiple Sclerosis	*DR 2*	*4.8*[b]
Ankylosing Spondylitis	*B 27*	*87.4*[b]
Rheumatoid Arthritis	*DR 4*	*5.8*[b]
Dermatitis Herpetiformis	*DR 3*	*56.4*[b]

[a]*Svejgaard et al. (1980)*
[b]*Ryder et al. (1979)*

Relative Risk (RR) = ad/bc

a = number of individuals with the disease and with the antigen
b = number of individuals with the disease and without the antigen
c = number of individuals without the disease and with the antigen
d = number of individuals without the disease and without the antigen

Therefore the relative risk is not an estimate of how much more frequent an antigen is in the diseased group compared to the disease-free group.

However, in future it is probable that those wishing to breed for disease resistance will have a number of specific resistance genes to work with. Much work has been devoted to the detection of disease resistance genes in certain diseases. For example, in humans, resistance to malaria is affected by the Duffy red blood cell system (Miller *et al.*, 1975, 1976), the Haemoglobin system (Allison, 1954), and possibly the Glucose 6-phosphate dehydrogenase system (Motulsky, 1964, 1975) and human MHS (Piazza *et al.*, 1973). Recent reviews of work in humans (Mourant *et al.*, 1977) and mice (Skamene *et al.*, 1980) highlight many exciting prospects in this area.

Another way of breeding for disease resistance exists. It is possible to measure immune responsiveness without exposing an animal to any risk of disease. This can be done by immunization with a harmless antigen such as xenogeneic red blood cells (Biozzi *et al.*, 1975; Siegel and Gross, 1980; Pevzner *et al.*, 1981) or by measuring immune responses *in vitro* (Newman *et al.*, 1977). As the immune response is a good, but not infallible, indicator of resistance to infection, these procedures may be useful in breeding for disease resistance. In future it will be possible to improve disease resistance either by increasing the frequency of specific genes or by the more traditional method of selecting for improvements in a multifactorial trait (immune responsiveness).

However, disease resistance is only one of many desirable traits. Animals which are poor producers will not be generally accepted even if their disease resistance is excellent. Resistance controlled by detectable genes will probably be easier to combine with other desirable characters in a breeding programme.

One concern in breeding for disease resistance is the possibility of resistance breakdown. In plant breeding, the resistance of some plant varieties to certain diseases has broken down as new strains of pathogen have evolved. We do not know if a similar fate would overtake animals bred for disease resistance. More work in this area is needed. Russell (1978) has argued that the problems of resistance breakdown have been exaggerated. There have been very few catastrophic pest or disease attacks as a result of breakdown of resistance. Resistance breakdown has been more common with single genes giving total or near total resistance by the mechanism of hypersensitivity. However, in animals resistance is not due only to single genes. Although single genes associated with resistance can be identified, they do not usually confer total resistance to disease. Even the Duffy-null allele which confers total resistance to *Plasmodium vivax* malaria does not confer resistance to *Plasmodium falciparum* malaria (Miller *et al.*, 1976). In plants, polygenic resistance or resistance controlled by major genes with modifiers has usually been durable (Russell, 1978). Also, in animals the evidence suggests that both pathogens and host adapt to each other to produce a less severe disease (Smith, 1934; Cockburn, 1963; Burnet and White, 1972). This is reasonable as a pathogen which rapidly kills its host will usually be less able to propagate than a pathogen from another strain which produces a less severe disease. For example, myxoma virus in American rabbits is virtually symptomless and Australian rabbits and the myxoma virus appear to be adapting to each other to produce a less severe disease (Rendel, 1971).

The desirability of breeding for disease resistance will vary with the disease and the effectiveness of other control measures. It may be that a combination of measures will be most useful. In chickens, both vaccination and selective breeding are necessary to maximise resistance to Marek's disease (Gavora and Spencer, 1979). A combination of different methods is more likely to provide durable disease control (Russell, 1978).

II Better Characterisation of Genetic Variation

Another important consideration for animal breeders comes from the fact that the explosive growth in both molecular and immunogenetics will mean better characterisation of genetic variation. Already, many enthusiasts are talking about identifying all the genes of the human genome. Even if this ideal is not achieved there will be a tremendous spin-off for animal breeders. Many of the laboratory methods used to detect human genes can be used with minor variations to detect genes in other species. Better detection of parentage errors will become possible. An improved ability to detect genetic variants in domestic livestock will allow us to determine more accurately which breeds are closely related by descent (Rendel, 1967). This may allow us to determine which breeds are best crossed to maximise heterosis for a given trait.

There is another possibility which deserves attention. Current dairy cattle breeding practice means that a small number of sires (through artificial insemination) are transmitting their genes to a large number of offspring. The possibility exists that a deleterious gene could achieve a high frequency in dairy herds. If the deleterious gene is recessive, natural selection will take a long time to reduce the gene frequency. However, it may become possible to detect the deleterious gene in the heterozygous state which would be of considerable advantage.

ACKNOWLEDGMENTS

I thank Professor B. Morris, Dr. F. Nicholas, Dr. S.W. Serjeantson, Mr. P.M. Summers and Mr. A.R. Hugh for their many helpful comments. The responsibility for the ideas expressed in this paper is mine.

REFERENCES

Allison, A.C. (1954). *Br. med. J. 1,* 290.
Benacerraf, B. (1981). *Science 212,* 1229
Biozzi, G., Stiffel, C., Mouton, D., and Bouthillier, Y. (1975). *In* "Immunogenetics and Immunodeficiency" (B. Benacerraf, ed.), p. 179. MTP, Lancaster.
Burnet, F.M., and White, D.O. (1972). "Natural History of Infectious Disease." Cambridge University Press, Cambridge.
Cameron, H.S., Hughes, E.H., and Gregory, P.W. (1942). *J. Anim. Sci. 1,* 106.
Cockburn, A. (1963). "The Evolution and Eradication of Infectious Diseases." Johns Hopkins Press, Baltimore.
Dausset, J., and Svejgaard, A. (1977). "HLA and Disease." Inserm, Paris.
Dungal, N., Gislason, G., and Taylor, G.L. (1938). *J. comp. Path. Ther. 51,* 46.
Fortner, J. (1949). *Berl. Münch. tierärztl. Wschr.* 37.
Gasser, D.L., Newlin, C.M., Palm, J., and Gonatas, N.K. (1973). *Science 181,* 872.
Gavora, J.S., and Spencer, J.L. (1979). *Comp. Immunol. Microbiol. Infect. Dis. 2,* 359.
Götze, D. (1977). "The Major Histocompatibility System in Man and Animals." Springer-Verlag, Berlin.
Hutt, F.B. (1958). "Genetic Resistance to Disease in Domestic Animals." Comstock Publishing Associates, Ithaca, New York.
Lambert, W.V. (1932). *J. Immun. 23,* 229.
Legates, J.E., and Grinnells, C.D. (1952). *J. Dairy Sci. 35,* 829.
Lilley, F. (1966). *Genetics 53,* 529.
Longenecker, B.M., and Mosman, T.R. (1981). *Immunogenetics 13,* 1.
Miller, L.H., Mason, S.J., Dvorak, J.A., McGinnis, M.H., and Rothman, I.K. (1975). *Science 189,* 561.
Miller, L.H., Mason, S.J., Clyde, D.F., and McGinnis, M.H. (1976). *New Engl. J. Med. 295,* 302.
Motulsky, A.G. (1964). *Am. J. trop. Med. Hyg. 13,* 147.
Motulsky, A.G. (1975). *In* "The Role of Natural Selection in Human Evolution" (F. Salzano, ed.), p. 271. North-Holland, Amsterdam.
Mourant, A.E., Kopec, A.C., and Domaniewska-Sobczak, K. (1977). "Blood Groups and Diseases." Oxford University Press, Oxford.

Newman, W., Stoner, G.L., and Bloom, B.R. (1977). *Nature (Lond.) 269*, 153.
Oldstone, M.B.A., Dixon, F.J., Mitchell, G.F., and McDevitt, H.O. (1973). *J. exp. Med. 137*, 1201.
Pevzner, I.Y., Stone, H.A., and Nordskog, A.W. (1981). *Poult. Sci. 60*, 920.
Piazza, A., Belvedere, M.C., Bernoco, D., Conighi, C., Contu, L., Curtoni, E.S., Mattiuz, P.L., Mayr, W., Richiardi, P., Scudeller, G., and Ceppellini, R. (1973). *In* "Histocompatibility Testing 1972" (J. Dausset and J. Colombani, eds.), p. 73. Munksgaard, Copenhagen.
Rendel, J. (1967). *Anim. Breed. Abstr. 35*, 371.
Rendel, J.M. (1971). *Search 2*, 89.
Roberts, E., and Card, L.E. (1935). *Ill. Agric. Exp. Stn. Bull. 419*, cited by Hutt, F.B. "Genetic Resistance to Disease in Domestic Animals." Comstock Publishing Associates, Ithaca, New York.
Robertson, A., and Barker, J.S.F. (1966). *Anim. Prod. 8*, 241.
Rosenberg, M.M., Alicata, J.E., and Palafox, A.L. (1954). *Poult. Sci. 33*, 972.
Russell, G.E. (1978). "Plant Breeding for Pest and Disease Resistance." Butterworths, London.
Ryder, L.P., Andersen, E., and Svejgaard, A. (eds.) (1979). "HLA and Disease Registry. Third Report." Munksgaard, Copenhagen.
Sellwood, R. (1979). *Vet. Rec. 105*, 228.
Siegel, P.B., and Gross, W.B. (1980). *Poult. Sci. 59*, 1.
Skamene, E., Kongshavn, P.L., and Landy, M. (eds.) (1980). "Genetic Control of Natural Resistance to Infection and Malignancy." Academic Press, New York.
Smith, T. (1934). "Parasitism and Disease." Princeton University Press, Princeton.
Spooner, R.L., Bradley, J.S., and Young, G.B. (1975). *Vet. Rec. 97*, 125.
Svejgaard, A., Platz, P., and Ryder, L.P. (1980). *In* "Histocompatibility Testing 1980" (P.I. Terasaki, ed.), p. 638. University of California at Los Angeles Tissue Typing Laboratory, Los Angeles.
Utech, K.B.W., Seifert, G.W., and Wharton, R.H. (1978). *Aust. J. agric. Res. 29*, 411.
Ward, A.H., and Castle, O.M. (1945). *New Zealand Dairy Board. 21st Ann. Report*, p. 46.
Warwick, B.L., Berry, R.O., Turk, R.D., and Morgan, C.O. (1949). *J. Anim. Sci. 8*, 609.
Windon, R.G., and Duneen, J.K. (1981). *Int. J. Parasitol. 11*, 11.
Williams, R.M., and Moore, M.J. (1973). *J. exp. Med. 138*, 775.

Discussion Summary

While breeding for disease resistance is unlikely to be the optimum method for reducing production losses for all diseases affecting economic livestock, genetic control of disease could well become a more important component of future breeding programs. However, it is clear that many uncertainties need to be resolved before the potential of genetic control of disease can be evaluated. The primary problems identified relate to (a) the equilibrium between host and pathogen, and (b) the genetic mechanisms of resistance.

The equilibrium between host and pathogen involves two components, one in terms of numbers of pathogens which is not independent of the second, the genetic equilibrium. Clarke (1976) has examined some aspects of the demography and genetics of host-parasite systems, but further modelling and theoretical studies are necessary.

While it is recognized that in natural populations, a pathogen may evolve towards lower virulence (as in the case of the myxoma virus in Australian wild rabbits), this may have little relevance to predicting genetic changes in pathogens when the host is artificially selected for increased resistance. Mutation will be injecting new genetic variation into both host and pathogen, and it is commonly supposed that a pathogen may evolve rapidly to overcome host resistance, both because of larger population sizes and shorter generation interval. However, the pathogen does not always win in its battle to survive. In some cases, it may be easier for the host to change the nature of its resistance or tolerance, than it is for the pathogen to change its mechanism of pathogenicity. Identification of particular host-pathogen systems where this might be so would be extremely valuable. This possibility will depend on the genetic nature of resistance (single gene or polygenic) and on the resistance mechanism.

Few, if any, generalizations yet appear to be possible with regard to the genetic basis of resistance. It was suggested that resistance to metazoan parasites may have a more complex basis than resistance to bacteria and viruses, but no supporting evidence could be cited. The speaker had emphasised that at least for many diseases studied in man, resistance is associated with the presence of a single allele. The exploding study of the major histocompatibility system (MHS) in man and its association with a variety of diseases may point the way to future developments in livestock species.

Where host resistance is not immunologically mediated, the resistance mechanism may be complex and under polygenic control. In such cases, selection to increase host resistance could produce a more stable system, i.e., one less likely to break down because of genetic changes in the pathogen.

While immunological studies, in particular of the MHS in domestic livestock species should be pursued vigorously, quantitative genetic analysis also is appropriate. Where little is known about a particular host-pathogen or host-parasite system, divergent selection should be used to produce lines having high and low resistance, so that the mechanism of resistance can be studied.

Whether it is possible, and if so, how to breed for more or less general disease resistance in farm animals has yet to be determined.

REFERENCE

Clarke, B. (1976). *In* "Genetic Aspects of Host-Parasite Relationships" (A.E.R. Taylor and R. Muller, eds.), p. 87. Blackwell, Oxford.

Part 4
Reproductive Biology

Further Possibilities for Manipulating the Reproductive Process

R. B. Land
I. K. Gauld
G. J. Lee
R. Webb

ARC Animal Breeding Research Organisation
West Mains Road
Edinburgh

The incentive to improve reproduction is considerable. So too is the opportunity : there is genetic variation; physiological measures of that variation are potentially available; such measures would increase the rates of response; only when cattle produce quads will their genetic superiority for reproduction equal the present genetic superiority of the dairy cow over the expected milk yield for a mammal of their size.

Variation in the activity of the ovary is largely accepted as the principal component of variation in the fecundity of cattle and sheep and hence as the key to genetic improvement. Intensive studies of the physiological control of the ovary in recent years and of genetic variation in that control have not however led to a clear indication of appropriate physiological criteria of fecundity. Many aspects of ovarian activity are postulated to be related to the sensitivity of gonadotrophin release to the negative feedback effects of gonadal hormones, but this is difficult to measure directly and fails in the sense that it is an operational rather than a functional description. When ovulation rate and seasonality are genetically independent, they must affect negative feedback sensitivity in a different way but the physiological separation of these is not understood.

Physiological understanding is still at the stage that the number of options as criteria for selection is increasing rather than decreasing.

FUTURE DEVELOPMENTS IN THE
GENETIC IMPROVEMENT OF ANIMALS
ISBN 0 12 078830 6

It is suggested that establishment of replicated populations selected for specific components of the reproductive process would markedly aid the identification of physiological criteria which are neither age nor sex limited. Such populations would test the importance of the component and if successful enable potential criteria to be assessed in no more time than that needed to take the measurement itself. The alternative, to study unselected pedigree populations, has the drawback that there is a considerable time lag between the realisation of a potential criterion and the evaluation of its worth; divergent selection for a possible criterion is efficient for that one but again is slow and has to be repeated for each new criterion or refinement.

Possible components would be ovulation rate in sheep and cattle, aseasonality in sheep and embryo survival in cattle. Physiology may aid genetic selection but equally genetic selection may aid the identification of key physiological components.

I Introduction

The very title of this paper is an invitation to speculate and an approval for licence. Speculation is however only useful within a limited framework and here the main limits are time-scale and finance. In deference to reason therefore, it is proposed to accept the limits of what in terms of the genetics of large animals is the relatively short time period of the next 25 years or so and to assume that even an increasingly affluent and hungry world will not greatly increase the resources allocated to improve the supply of animal product.

The identification of further possibilities for genetic manipulation then falls into two stages. It is first a question of what restricts the reproductive process and secondly of what are the genetic options to alleviate those restrictions. While the precise answers to these questions would be expected to be species specific, the nature of the answers might well be similar for cattle, goats and sheep on the one hand and pigs and rabbits on the other. In an Australian context it is only proposed to consider cattle and sheep. Even then, most research has been conducted on sheep and conclusions for cattle are often dependent upon extrapolation and inference. Poultry are not considered.

"Further possibilities" beg an origin but the Congress on Sheep and Cattle Breeding held in New Zealand makes 1980 a practical and recent base. Reviews and papers by Bindon and Piper (1982), Bradford, Barkley and Spearow (1980), Hanrahan and Quirke (1982), Land (1982) and Piper and Bindon (1982) are appropriate entries to the literature. It is not intended to be comprehensive but rather to consider possibilities which show promise. Even then, the accuracy of judgement clearly decreases as one approaches the end of the 25 years.

Finally but perhaps the most important starting point is the knowledge that sheep (and mouse) fecundity responds to genetic selection. Increases in the rate of response and in convenience of selection will be given particular attention.

II The Possibilities in 1980

A. Genetic Variation

A range of options is open for genetic manipulation of the reproductive performance of sheep. Litter size varies among breeds from one to three lambs, the superiority of the Finnish Landrace and Romanov being largely inherited additively; current evidence shows the superiority of the Booroola Merino to arise from a single allele with a dominant effect of two or so lambs on litter size; and, selection within populations has been shown to change fecundity by 15% or so in 10 years. Sheep may give birth to six lambs. The duration of post-partum and seasonal anoestrus varies among breeds; genetic variation in the duration of seasonal anoestrus has been reported within them (Purser, 1973).

For cattle, the immediate options are limited. There is little variation among breeds and realised heritabilities have not been estimated for any reproductive traits. Genetic variation within populations is only now being studied by direct selection but extrapolation from the results of genetic selection in other mammalian species and the estimates of the heritability of twinning in cattle indicate that fecundity would respond to selection.

In general prolificacy but not fertility has been found to be heritable. To the extent that fertility represents "successful" ova, however, a conspicuous exception to this generalisation is the high embryo survival line of mice selected by G.E. Bradford. With survival rates of 90% as compared with 75% in control populations and 50% in other selected lines, Bradford showed that the 30% or so embryo mortality common among mammals is not obligatory. Favourable recessive alleles for survival were segregating in his base population.

For cattle, and for sheep once the possibilities of population substitution have been exhausted, improvement is dependent upon genetic selection within populations. What opportunities are there to increase the rate of response to selection?

B. *The Measurement of Superiority*

The measurement of reproductive superiority is restricted by the sex and age limited expression of most of the component traits and by the high contribution of environmental variation to the determination of the phenotype. There is consensus that the establishment of techniques to improve the measurement of genetic superiority would have a marked impact on the rates of response and on the ease of selection for reproductive traits.

Walkley and Smith (1980) have recently quantified the advantages of criteria of fecundity which can be measured in both sexes before puberty. The use of a criterion with a genetic correlation with fecundity of 0.5 and heritability of 0.35, in combination with fecundity itself would increase the rate of response in fecundity to almost twice that possible with selection for fecundity alone. More importantly, the use of such a trait for selection among males alone would give a rate of response a little greater than would selection on fecundity itself without any need to record fecundity.

Criteria suggested up to this time have been based on the hypothesis that ovarian activity is an underlying limit to the reproductive performance of sheep and cattle, that elements in the endocrine control of the gonads are common to males and females and that the genes which influence these common elements may be expressed before puberty. There is then the possibility of using a measure of FSH and LH to form a criterion, or of using testis size as an integrator of gonadotrophin activity over time. Of these possibilities, FSH has not been evaluated; LH has been shown to have a low repeatability, although measurement following an LH-RH stimulation test may have a higher repeatability; there are some reports of a positive genetic correlation between testis size and ovulation rate. Before discussing these criteria, their improvement and possible alternatives, it is proposed to consider the current ideas of the endocrine control of ovarian activity.

III The Endocrine Control of Ovarian Activity

What controls the *number of eggs shed*? Evidence points to the importance of the balance between the stimulatory effects of gonadotrophic hormones and the feedback effects of gonadal hormones on the inhibition of gonadotrophin release. The high ovulation rate of Finnish Landrace and Romanov sheep is thought to arise at least in part from an ability to continue the secretion of trophic hormones in the presence of higher concentrations of feedback hormone from the ovary. The Booroola gene is thought to reduce the production of feedback hormones per egg produced; Bindon *et al.* (1981) report the concentration of progesterone in peripheral plasma on Day 11 of the oestrous cycle to be the same in ewes shedding one to eight eggs.

Any gene which reduces steroid and inhibin production per follicle, which decreases their biological half-life, which decreases their perception by the hypothalamus and pituitary gland or which decreases the response of these organs may therefore be a gene for higher ovulation. These possibilities are illustrated in Figure 1.

The duration of the *breeding season* is thought to be determined by seasonal variation in the ability of the hypothalamus and pituitary gland to maintain the pulses of LH at an adequate frequency and amplitude for the maturation of developing follicles in the presence of concentrations of oestrogen typical of such pre-ovulatory follicles. On long days sheep are sensitive to oestrogen feedback and cannot maintain adequate pulses of LH.

The activity of the pineal gland may well influence the response of the hypothalamus, but the relationship is not yet well understood. The duration of secretion of melatonin by the pineal gland usually corresponds to the duration of darkness. Removal of the gland 'allows' ovulation to continue on long days (Bittman and Karsch, 1981). The infusion of melatonin to pinealectomised sheep each day for a period of time equal to the duration of a short night induces sensitivity to oestrogen and blocks ovulation; the addition of melatonin before darkness to an intact sheep on long days to correspond to the duration of a long night may lead to the resumption of reproductive activity.

Part of the difficulty in understanding the control of anoestrus could arise from its apparent subdivision into two components, an intrinsic rhythm of reproductive activity which groups the oestrous cycles and the induction of that rhythm by changes in the photoperiod. In the absence of cues

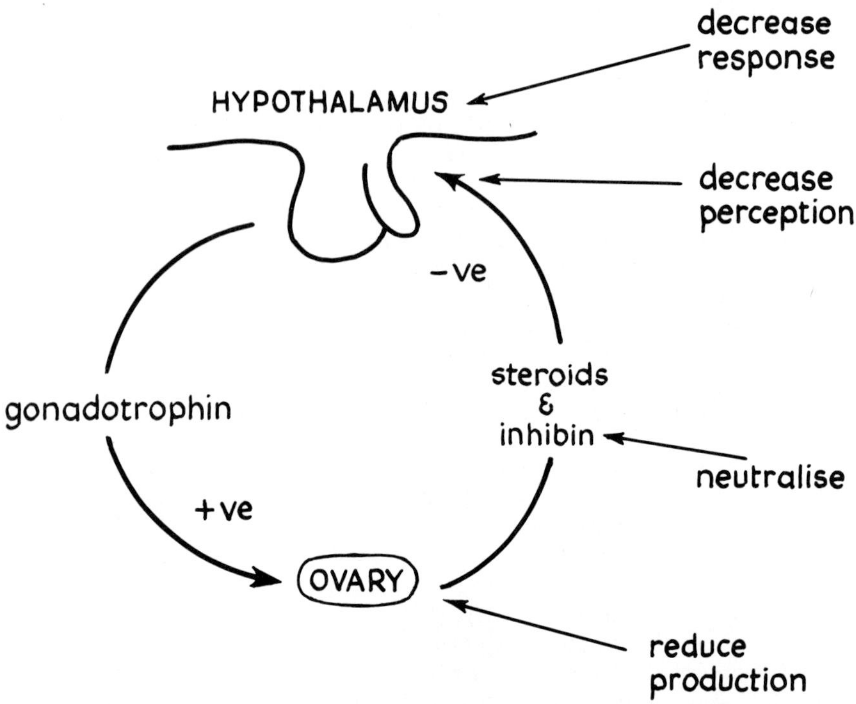

FIGURE 1. The feedback equilibrium controlling the number of eggs shed and examples of steps where genes may act to decrease negative feedback and hence increase ovulation rate.

from the photoperiod, ewes tend to alternate between periods of acyclicity and a series of consecutive oestrous cycles. Photoperiodic cues act on this intrinsic rhythm to coordinate ewes within a group. Melatonin however is not the whole answer; work in Edinburgh shows that rams kept on constant daylength - long or short - may spontaneously grow and regress their testes in synchrony with changes in the pattern of melatonin secretion (O. Almeida, personal communication). The control of the grouping of oestrous cycles in the absence of photoperiodic cues is not known.

The role of prolactin in seasonal anoestrus is now thought to be slight. Although a period of elevated prolactin corresponds to the period of anoestrus in some breeds of sheep, others return to cyclicity while prolactin is still high, while ovarian cycles continue in others after the long day rise in prolactin (G.M. Webster and W. Haresign, personal communication, 1981; Carr and Land, 1982 respectively). The suppression of prolactin with CB154 does not increase sensitivity to positive feedback, or enable those follicles which ovulate in response to LH-RH to form functional corpora lutea (see Land *et al.*, 1980). By contrast the recent demonstration of an antagonistic effect of prolactin on follicular function indicates its role in the control of ovulation rate has yet to be appreciated.

Post-partum anoestrus in sheep and cattle is again associated with inadequate gonadotrophin support for follicular development and high sensitivity to gonadal hormone feedback. Although it is less well understood than seasonal anoestrus in sheep, it again is thought to be functionally independent of the associated higher concentration of prolactin in peripheral plasma.

Finally, also *puberty* is the eventual persistence of gonadotrophin release in the presence of a functional gonad (Ryan and Foster, 1980).

Figure 2 illustrates some relationships between LH pulse frequency and gonadal function.

The ability to establish and then to maintain gonadotrophin secretion in the presence of a high ovulation rate, shortly after parturition, at an earlier age than normal and, for sheep, in long days is then a common feature of superiority for all four aspects of ovarian activity. The specificity of the feedback factors to these four activities has yet to be elucidated: it is possible that alleles may segregate for persistent gonadotrophin release regardless of the circumstances as well as for each of the specific causes of insensitivity which underlie the independent traits. Fortunately there are some clues. In particular follicle development but not the number of follicles which develop seems to be dependent upon the rate of LH pulses - above a certain rate they develop, below it they do not.

At what level might genetic superiority be measured?

IV Levels of Measurement

The aim is to identify the presence of favourable alleles either with greater precision than is possible by measuring the trait itself or in circumstances in which the trait is not normally expressed; for example litter size in a young male. What possibilities might be available?

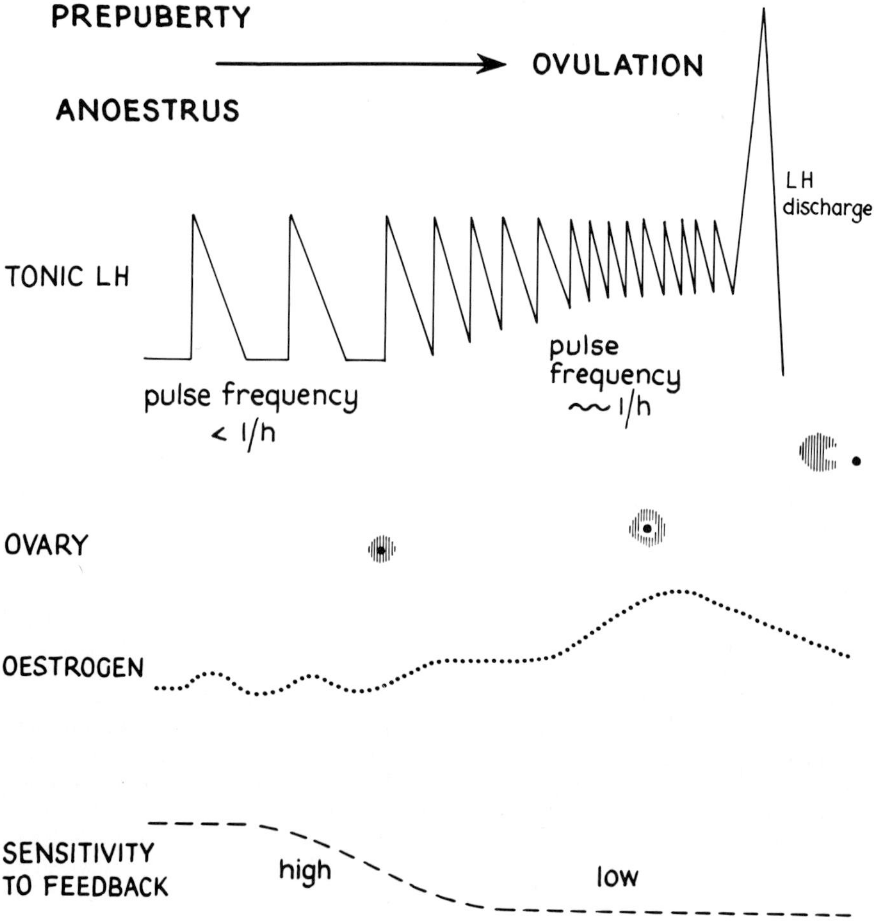

FIGURE 2. Illustration of the increase in tonic LH pulse frequency (No. per hour) through to the discharge of LH, the rise in oestrogen and ovulation as the sensitivity of the release of gonadotrophins to the negative feedback effects of gonadal steroids decreases (after Ryan and Foster, 1980).

The nearer one approaches the gene itself the greater the opportunity for direct manipulation. With, say, a complete restriction enzyme map for the species in question and knowledge of the alleles which were desirable it would be possible to manipulate minor genes by 'Mendelian Genetics', but individual favourable alleles still have to be identified. Further, the cost might well be too high to be acceptable.

Kacser and Burns (1979) present a consideration of the possibility that individual genes may limit performance. They base their arguments on the theory that genes are expressed via enzymes which in turn control the flux through a biochemical system. The sensitivity coefficient describes the effect on the flux of a given change in the activity of an enzyme. This declines from one to zero as the activity of an enzyme is increased; for a system the sum of these coefficients is unity. If the activity of one enzyme in a pathway is increased, the sensitivity coefficients of other enzymes in that pathway will inevitably increase as that of the enzyme concerned decreases. If there are few loci controlling a flux which has a high genetic variance and a high correlation with a trait of commercial importance, it would be useful to be able to measure the individual genes but these are the very circumstances in which the influence of a gene might be sufficiently large for its effects on the trait concerned to be identified directly (Fig. 3). On this argument it was considered that a physiological level may be more appropriate than a biochemical level for the measurement of the presence of favourable genes in the short term (Land, 1981). There may, however, be exceptions to this generalisation and biochemical measurement may have a role to play, especially in particular circumstances. With a dominant allele of large effect such as the postulated Booroola gene, for example, it would be useful to be able to discriminate between heterozygotes and homozygotes, and if the product of

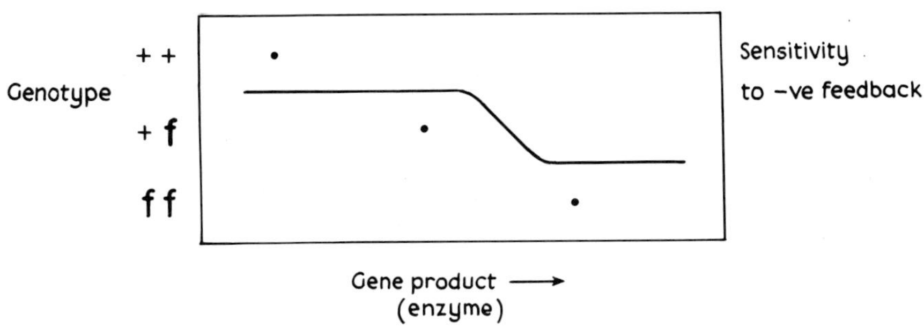

FIGURE 3. Example of a hypothetical gene (f) which increases gene product activity (●) but has a recessive action on fecundity through its control of the sensitivity of gonadotrophin release to the negative feedback effects of gonadal hormones (——) (after the dominance model of Kacser and Burns, 1981).

the gene concerned could be identified and measured, this could be done. If this could be assessed in males, the benefits would be even greater.

Dominant genes, however, can be identified in the phenotype if their effects are large enough; and they can be introduced to other populations by a combination of backcrossing and progeny testing. Recessive genes might never be detected and hence could be important candidates for biochemical detection (e.g., Fig. 4). The likelihood of a favourable recessive mutant becoming homozygous is low. The likelihood of two such animals being mated together for the inheritance of their superiority to be recognised is even less - especially for genes controlling the reproductive process when merit cannot normally be recognised in both sexes, assortative mating is not possible, and environmental variation is high. If the Booroola gene had been recessive it is much less likely that it would have been identified and maintained.

The logical extension of this argument is to say that if we knew the sensitivity coefficients of all of the enzymes involved in the control of the reproductive process, we could screen large populations for the presence of favourable recessives. This, however, is somewhat unrealistic. A first step might be to say that if we knew the primary gene product of the Booroola gene we could screen, say cattle to pick up recessive alleles with the same effect.

A second approach might be to intensify current efforts to identify the limits to reproduction in physiological terms, study the biochemical basis of those physiological characteristics and then move to identify alleles with favourable effects. The use of biochemical measurement would then be a sophistication of an initially proven physiological criterion and a foreseeable step in a logical sequence of study.

The present case for the use of biochemical measures of gene action rests largely on the wish to screen for recessive alleles with large effects and to discriminate between heterozygous and homozgous dominants in both sexes. It is therefore very difficult to evaluate the merits of a biochemical approach. With a large effect it could be argued that one only has to identify one allele for the case to be made and the problem solved; the likelihood of finding that allele, however, is low. The stakes are high but limited investment in such a programme would seem appropriate.

The Booroola gene is not the only example of a single gene with a major effect, sex reversal (Sxr) and testicular feminisation (Tfm) are two dramatic examples which have been reported in more than one species. The hypogonadal gene

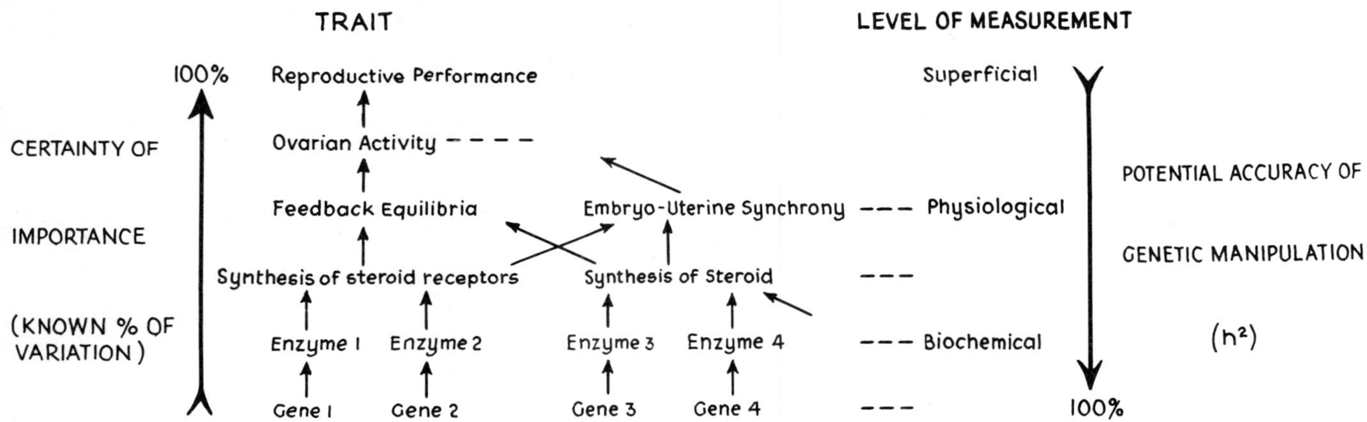

FIGURE 4. Illustration of the balance between the advantages and drawbacks of various levels of measurement of genetic superiority. The dotted lines acknowledge the limits to comprehension.

(hpg) is found only in mice so far and there are other single genes such as obese (ob) and dwarf (dw), polling in goats and double muscling in cattle that affect reproduction as well as stature. Of these, Tfm acts to block the response of target tissues to the masculinisation of testosterone; in terms of gene action it might block the induction of an enzyme. The possibility that other single genes may have a major effect, for example, on the detection of feedback by the hypothalamus is very real but we need to know how the hypothalamus acts before we can start to screen for advantageous recessives.

There are then specific circumstances in which a biochemical level of measurement of the presence of favourable alleles may be appropriate. In the long term such an approach would link through to the biochemistry of gene action and to the mapping of the genome to give what with our present limited horizons may be thought of as the ultimate solution of direct genetic manipulation.

In the meantime there is the possibility to use a physiological level of measurement. This may increase rates of response to selection and allow traits to be recognised more precisely so that unfavourable correlations may be split. Rather than try to identify the presence of individual favourable alleles it may be possible to identify the presence of several at the same time through their effect on a physiological system. As a simple example: If the concentration of FSH in peripheral plasma were found to be important in particular circumstances, it would be simpler to measure that concentration than to measure the individual presence of alleles which had a favourable effect on reproduction through their contribution to FSH synthesis, release or catabolism. The rationale would now be that the physiological system would be closer to the expression of the trait itself, it would be expected to have a high 'sensitivity' coefficient and that the regression of the trait on the activity of the physiological system would be expected to be linear for a greater change in the physiological trait than for individual enzymes (Fig. 4).

V Studied Physiological Criteria of Ovarian Activity

The predominant role of variation in ovarian activity in the control of variation in prolificacy suggests the use of measures of the physiological control of the ovary as criteria for selection in young animals of both sexes. This possibility has been studied during the past 10 years or so; what further possibilities arise from the results of these studies?

Two programmes of selection have been conducted in sheep where a physiological characteristic has been used as the criterion for selection. Both have been conducted at ABRO, neither has been shown to be a characteristic appropriate for use in the industry, but the results do indicate the nature of the mistakes that have been made, the way in which they might be corrected and, of particular importance, the questions which should be asked.

A. Testis Size

In one experiment divergent lines were selected on the basis of the diameter of the testis at around 10 weeks of age, corrected for body weight. The aim was to change ovulation rate without changing body weight but exactly the opposite has happened, ovulation rate has remained relatively constant but the mature females of the low line are heavier (approximately 10 kg) than those in the high line. A second change is the earlier onset (approximately two weeks) of the breeding season in the line selected for large testes.

The correlated change in the time of end of anoestrus could have arisen if testis size during the summer indicates the presence of genes for sensitivity of reproduction to changes in the photoperiod. To use testis size as a criterion of genes for ovulation rate measurement should therefore possibly be deferred to a later age when the males are in as near as possible the same environment as that in which the females reproduce.

The body weight change might be both more complex and more interesting. We could simply have malcorrected: The correction was on the basis of body weight at the time of testis measurement and that weight has hardly changed; adult body weight which we wished to hold constant was not of course known at the time of selection. Alternatively we could have found a method of selection to change growth curves; higher testosterone concentrations from the large testes terminating body growth earlier. The relationship between growth and reproduction will be referred to again later.

A further result in the sheep testis selection experiment is an indication that conception rates are falling in the low line but as matings have always been conducted within lines this could be a ram effect. Regardless of the involvement of either or both sexes, this possible change illustrates the difficulty of interpreting divergent selection lines without controls, despite their greater efficiency. We do not know whether the low line deteriorated or the high line improved.

On the basis of these results a new experiment has been established where the effects of selection for large testes in lambs of 20 weeks (corrected only for birth type) is to be assessed by comparison with an unselected control line. (The use of a control rather than a downward selection line is indicated in this experiment also by the difficulty of avoiding males with small testes for unusual reasons, e.g., runts.)

The present change in the ratio of ovulation rate to body weight, published material quoted in the papers mentioned earlier, recent laboratory animal data (Eisen and Johnson, 1982), and unpublished reports of a relationship between testis size and ovulation rate in Norway (T. Stein, personal communication), Western Australia (B.R. Beetson and C.A. Morris, personal communication) and Ireland (J.P. Hanrahan and J.F. Quirke, personal communication), all indicate the merits of further study. The sort of questions which now need to be asked are the type of environment in which testis size should be measured (age and season), the importance of body weight and the generality of answers to different breeds and latitudes.

B. *The LH Response to LH-RH*

The episodic release of LH and consequent low repeatability of a single measurement of the concentration of LH in peripheral plasma is a strong contra-indication to the use of tonic LH concentrations as a criterion for selection. One way to overcome this is to use the concentration after LH-RH, which has a higher repeatability.

The direct response to selection for the log mean concentration of LH, 30, 50 and 70 minutes after the intravenous injection of 5 μg LH-RH in terms of the mean concentration after injection is:

Years of selection among males	Selection line High	Selection line Low	Ratio (High/Low)
0	30.2		
1	48.6	25.8	1.9
2	34.3	19.9	1.7
3	31.0	14.4	2.2

It is too early to assess correlated responses as the standard errors are very large, but the female characteristics after one and two years of selection among males are:

	Years of selection			
	1		2	
Line	High	Low	High	Low
1st season				
No. of females	41	43	51	52
Median date of 1st oestrus (day)	326	319	324	329
Per cent mated	100	100	100	96
Per cent lambed	75	87	57	60
No. of lambs born per ewe alive	1.34	1.37	0.96	0.94
No. of lambs born per ewe lambing	1.78	1.56	1.69	1.57
2nd season				
No. of females	39	37		
Median date of 1st oestrus	266	245		
Per cent mated	100	100		
Per cent lambed	72	86		
No. of lambs born per ewe alive	1.26	1.73		
No. of lambs born per ewe lambing	1.75	2.00		

The large direct responses in the concentration of LH after LH-RH will give a powerful test of the biological relevance of this characteristic.

VI Potential Physiological Criteria

The consideration of recent advances in knowledge of the physiological control of the ovary, genetic variation in reproductive performance, and the information from physiological criteria studied so far are all relevant to the development of ideas for further manipulation of the reproductive process.

A. Whither LH?

LH has been the subject of extensive study because it is known to be involved in the control of the ovary, and assays for measurement in peripheral plasma were first developed for

LH. There are many data to show that LH is involved in the occurrence of ovulation and the priming of the follicle in preparation for the formation of a functional corpus luteum. As mentioned earlier, the pituitary gland must be able to maintain frequent (approximately hourly) pulses of LH in the presence of feedback hormones from developing follicles. There is, however, neither direct physiological evidence or physiological data from the comparison of sheep around the time of the ovulation of very different number of eggs to suggest that either the tonic concentration or the preovulatory discharge of LH relate in a simple direct way to the number of eggs shed. The preovulatory follicle develops receptors to LH and the number with receptors is similar to the ovulation rate of the breed (Webb and England, 1982), but LH has not been shown to change the number of follicles.

Selection to increase the reproductively active proportion of the lifespan is therefore likely to be selection which extends the conditions under which the hypothalamus and pituitary gland will maintain hourly pulses. Our problem at present is that this ability seems to be an all-or-none characteristic in the context of seasonal anoestrus. If oestrogen implants are used to simulate the presence of developing follicles, pulses of LH can be measured during the period of the normal breeding season but not during the period of normal anoestrus. The question then becomes "What determines the ability to maintain frequent pulses?". Potentially there is a continuously variable characteristic underlying the presently measured all-or-none trait but this has not yet been identified.

It is even possible that low LH levels might favour high ovulation rates. LH stimulates both oestrogen and progesterone production. Of these, oestrogen is principally responsible for the feedback control of LH during anoestrus while progesterone may principally control it during the breeding season (Karsch *et al.*, 1980). The lower the LH therefore the lower the steroidogenic stimulus to the developing follicle and hence the greater the number of follicles which might develop per unit of steroid production. Steroids also control FSH release and hence the lower the activity of LH required for follicle maturation the greater the potential to maintain FSH production. The observation of lower LH levels in Finn than Blackface (Land and Carr, 1979), in Romanov than Ile de France (Cahill *et al.*, 1981), and Finn than Suffolk (Webb and England, 1982) during the oestrous cycle may not be as incongruous as it might initially have appeared but rather be an indication of an alternative relationship between LH and ovulation rate.

B. Follicle Stimulating Hormone

The role of FSH in the control of ovarian activity and the measurement of genetic covariation with reproductive activity is a very neglected subject. Initially the hormone was difficult to measure but assays are now available, and of particular importance, it has a much higher repeatability than LH. A further pointer to the study of FSH is the observation from the data of Cahill *et al.* (1981) that the drop in the concentration of FSH from the luteal to the follicular phase in the Romanov was around 15% of the luteal level whereas in the Ile de France the drop from similar luteal levels was more than twice as great.

Normal levels of oestrogen and progesterone have been found not to suppress FSH to normal cyclic levels (Goodman *et al.*, 1981); non-steroidal components of follicular fluid (inhibin) are postulated to have a specific effect on FSH (see Franchimont and Channing, 1981 for reviews). Both pieces of information support the argument for the study of genetic covariation between FSH and ovarian activity.

C. Negative Feedback Control

The feedback control of FSH introduces the general question of the negative feedback control of ovarian stimulation. Luteal levels of progesterone and oestrogen will suppress LH to luteal levels although they will not suppress FSH. Romanov and Finnish Landrace sheep apparently maintain follicular development in the presence of higher concentrations of oestrogen. How this is mediated remains to be understood and only then can the possibility of separating the different component traits which act through feedback be assessed. Seasonality, puberty, post-partum fertility and ovulation rate are all to some extent at least genetically independent; variation must be mediated through different physiological pathways even though they have a common expression in the ability to generate frequent LH pulses in the presence of feedback hormones.

The location of the physiological determination of these traits within the hypothalamus and pituitary gland indicates that the variable underlying these characteristics is unlikely to be useful as a criterion for selection. The balance between gonadotrophin release and specific feedback hormones may be more revealing, but that presupposes knowledge of specific feedback hormones. Current evidence would suggest that pre-puberty, anoestrus and post-partum infertility may share oestrogen, while inhibin may be specific to ovulation rate.

Despite the deficiencies in understanding, it may be possible to think in terms of measuring the ability to generate hourly LH pulses or maintain the secretion of FSH in the presence of the appropriate feedback agent. In this way it would be possible to ensure the avoidance of, for example, animals with high LH as a result of gonadal quiescence. As discussed earlier, LH-RH may be a useful evocative agent to overcome the natural variation in tonic concentration so one might think of measuring the response to LH-RH in the presence of an oestrogen implant. The drawback is that if the response were particularly sensitive to the dose of feedback hormone the presence of endogenous hormones could still prejudice the superior animal.

Might the production of feedback hormones be an easier target? Much more is known about the biochemistry of oestrogen and progesterone production (although not inhibin) than feedback perception, and hence the opportunities to identify precise criteria are greater. Recent mouse data indicates genetic correlation between gonadal steroidogenesis in males and females and that this might arise from variation in the number of FSH binding sites (Wolfe *et al.*, 1981). Research might then investigate variation in LH induced androgen production by thecal cells, aromatisation of androgen to oestrogen by granulosa cells and the extent to which such variation may be common to males.

The pursuit of this approach is supported by the demonstration of the existence of alleles which affect neuraminidase activity and by the way in which FSH induces the aromatisation of androgens to oestrogens but that this in inhibited by prolactin (Wormack and Lu Shun Yan, 1981 and Dorrington and Gore-Langton, 1981 respectively). An allele for low neuraminidase could be an allele for high prolificacy through its effect on the number of sialic acid residues and hence the half-life of gonadotrophins. An allele for low aromatase activity either through insensitivity to FSH induction or sensitivity to prolactin inhibition would be an allele for high prolificacy. New research in ABRO is directed to study genetic variation in the steroidogenic pathways of the testis and ovary.

If the diversity of options were not sufficient there is limited data to indicate yet another problem. As with the possible negative correlation between LH and ovulation rate, there might be another paradox.

The introduction to intact Finnish Landrace ewes of a silastic implant of oestrogen, the release characteristics of which would be expected to maintain a concentration of oestrogen in peripheral plasma of approximately 4 pg/ml, depresses the ovulation rate. A second implant depresses it further. However, the ovulation rate increased from one oestrus cycle to the next when a half-sized implant was inserted alone:

No. of Animals with Change in Ovulation Rate

Change	-1	0	+1
Control	2	3	2
2 pg/ml	0	1	4
4 pg/ml	3	2	0

Equally, in an old experiment where it was shown that Finnish Landrace sheep are better able to withstand (in terms of ovulation rate) the daily injection of 25 or 50 μg oestradiol per day, the initial search for an appropriate dose again showed that small doses may increase the ovulation rate. Neither of these experiments alone is worthy of note but with the positive effect of oestrogen on follicular growth (e.g., Charlton *et al.*, 1982), they indicate a possible 'two-phase' action of oestrogen with an intermediate optimum.

We know that small titres of antibodies to oestrogens will increase the ovulation rate of sheep. Are we witnessing the reverse of the depression of ovulation rate in the presence of 4 or 8 pg/ml implants or the favourable effect of 2 pg/ml implants, i.e., do the antibodies reduce the availability; or do they extend the half-life of oestrogen and facilitate its favourable effect? Certainly similar titres do not affect the incidence of ovulation during anoestrus.

VII Embryo Mortality

The number of eggs shed might well be the major component of variation in fecundity, but 30% or so of eggs are lost. The mouse data referred to earlier shows that this is not inevitable.

Recent experiments in ABRO have shown that an asynchronous uterus may be deleterious rather than merely inadequate and this observation has led to the hypothesis that some embryonic mortality may arise from spontaneous asynchrony between mother and embryo (Wilmut and Sales, 1981 and unpublished). The timing of changes in uterine secretions is primarily dependent on changes in progesterone secretion and this is variable. Similarly there is variation in the stage of embryonic development. It may be that the advantages of the flexibility to reproduce in a variety of circumstances, inevitably leads to the drawback that reproduction will be less than ideal in any one environment.

If the duration of oestrus were a compromise between the benefits of a long duration to ensure mating and a short duration to ensure synchrony between embryo and uterus, natural selection could have actually favoured some embryo mortality as a by-product of the overall fitness of a female to produce offspring in a given environment. For example, a female with an oestrus period of 48 hours and 70% embryo survival would have greater fitness than a female with a 30 minute oestrus period and 100% embryo survival, in an environment where the mean time interval between the meeting of the sexes is two hours. Under such a hypothesis one can consider artificial selection for embryo survival with controlled conditions. This might be synchronisation of the time of ovulation with a progestagational or a luteolytic agent followed by insemination after a fixed interval. Genetic selection with a fixed time of insemination may lead to greater synchrony.

As with the physiology of negative feedback, one might be able to think in terms of enzymes or other proteins which are specifically deleterious to embryos at particular stages of pregnancy. Selection for the elimination or repression of such proteins could increase embryo survival. Once the importance of the presence or absence of single proteins is invoked, the possibility of single gene selection and biochemical screening rather than quantitative physiological selection should be considered. Indirectly this might return to a component of ovarian activity; steroids have 'essential' roles, including the 'preparation' of the uterus as well as their 'deleterious' roles maintaining low ovulation rates. Immunisation against a gonadal steroid increased the conception rate of a group of sheep mated at ABRO in 1980. Selection to change steroid production or to change steroid induction of gene action are complementary strategies.

The important point is to establish the idea that natural selection could favour intermediate embryo survival.

VIII Interactions with Health and Growth

Reproduction is just one component of the productivity of an animal and we have to ask whether changes in reproduction have or might affect other traits such as health, growth or lactation. The mouse data is particularly convincing; we saw earlier that selection for high embryo survival increased survival to 90% as compared with only 70% in a control population; the rate of survival in lines selected for other traits was only 50% (Bradford *et al.*, 1980). This was observed following selection for high growth as well as for low litter size.

The conventional way to approach the problem of unfavourable associations among traits is to use measures of genetic correlation and regression. If we are to move to the use of physiological criteria for selection, we can also consider approaching the problem from knowledge of the other physiological effects of those criteria. Although some reproductive hormones such as FSH are thought to be specific to reproduction, steroids for example have a known role in anabolism and in the case of prolactin its role in metabolism might be as important as its role in reproduction.

Much of the argument for possible covariation between prolactin and metabolism derives from the work of H.A. Tucker and colleagues at East Lansing (e.g., Peters *et al.*, 1978), who showed that daylength affects growth and lactation in cattle. Both these aspects of performance and the concentration of prolactin are increased on long days. The concentration of prolactin in peripheral plasma of sheep and deer changes in response to changes in the daylength. Quite dramatic effects have been observed in the efficiency of lamb growth. Lambs were found to eat 50% more food per unit of body growth when grown on short days than when grown on long days and to have lower prolactin concentrations (Forbes, 1975). The voluntary food intake of intact stags doubled on long relative to short days; in castrates the increase was less than 50% (Kay, 1979).

The relationship between prolactin, appetite and efficiency would have a real bearing on the use of selection criterion for aseasonality which incorporated a low concentration of prolactin. Selection for rapid lamb growth may have reinforced selection for seasonality of reproduction.

The dramatic reduction in the amplitude of seasonal variation in the appetite of deer in the absence of the gonads is an illustration of the way in which steroids may affect metabolism. In addition, males with testosterone grow faster and have leaner carcasses than females. High levels of steroids may terminate growth. Androgens depress wool production and the immune system; castration may improve wool production and the health of males.

This consideration of the interrelations between reproduction and other components of performance is intended to illustrate such possible relationships rather than be comprehensive. If we start to look at lactation for example the physiological interrelationships are even stronger. Qualitatively, reproduction is the usual prerequisite to lactation; within the process a simple example of a quantitative relationship between the two is the induction of greater milk yields by a large foetus, possibly via greater production of oestrogen or placental lactogen from the larger placenta. Conversely Berger *et al.* (1981) report a genetic basis to the antagonism between reproductive performance and yield of milk.

The more we understand of these relationships the easier it is to develop enlightened improvement programmes. This is particularly so if we start to use physiological criteria for selection.

If we drastically change seasonality, we may change efficiency. If we drastically reduced gonadal steroid output as a means to decrease negative feedback and hence raise prolificacy, we might also increase wool production but decrease early lean meat growth rates. Changing components of reproduction may not only affect other components, it might ultimately affect other traits. We can use both physiological and genetic manipulation to investigate these relationships. If we actively immunise against androstenedione to improve ovulation rate, do we increase steroid half-lives and improve efficiency, do we remove steroids and decrease efficiency or might the effects on the ovary be independent of changes in metabolic efficiency?

IX Strategy

The pursuit of suitable criteria for further genetic manipulation of the reproductive process is dependent upon five factors:

(1) A worthwhile aim
(2) Genetic variation
(3) A real advantage to be gained from using new criteria
(4) The possibility that such criteria might exist
(5) A suitable strategy of research.

Of these; reproduction is accepted to be a critical component of performance; genetic variation exists; Walkley and Smith (1980) showed the benefits of using a selection criterion that is not limited by sex or age; experiments to date have shown that new criteria may be there. What strategy is required?

The earlier discussion of possible physiological or biochemical criteria of reproductive performance deliberately emphasised the wide range of options and showed that current research strategies are leading to an increase rather than a decrease in those options. Physiological studies have tended to increase the number of elements in the qualitative description of the physiological control of reproduction. Genetic studies have been a mixture of comparisons among breeds and the occasional specific study of variation within breeds, either in terms of the segregation of a single gene or selection for a single possible criterion.

We are at the stage that we have enough knowledge to know what we want to know - the genetic covariation between the elements of control and quantitative measures of commercially important components of reproductive performance. What we need is a strategy of research to give us this information. Conventional physiological studies can continue to indicate further details of the elements of control, not the extent of genetic covariation. Equally, genetic studies among breeds will only tell us why those breeds differ. Single gene effects may be specific to those genes. Selection within particular populations for specific potential criteria gives little indication of the merits of other criteria. With such plethora of possibilities it is impractical to consider divergent selection for each possible trait, even though individually it may well be the most powerful genetic tool. Equally, it is impractical to consider the use of random bred pedigree populations for the measurement of all correlations between all traits; one might waste enormous resources measuring traits which are ultimately found to be useless. Further, whenever a new potential criterion is appreciated, it has to be incorporated into the recording system and the appropriate correlations calculated several years later. Breeds with differing prolificacy, seasonality, puberty, post-partum anoestrus should be studied in a single environment, populations should be selected for possible indirect criteria of merit and the effects of major genes should be studied in different backgrounds but replicated divergent populations would be an especially useful resource.

A. Replicate Divergent Populations

We already have considerable evidence to suggest the critical components of the reproductive process in cattle and sheep. Let us say conception rate and ovulation rate in cattle and ovulation rate and seasonality in sheep are the two most important components for each species. The strategy would then be to establish divergent populations for these traits. These populations could be established by 'up and down' selection, but as we are ultimately interested in improvement, 'upward' selection and the maintenance of random bred controls might be preferred; such an approach would also avoid the need to maintain unproductive 'down' lines; controls could be shared among traits. Such lines would confirm the feasibility of direct selection and once established, new indirect criteria could be evaluated immediately. To avoid being misled by the effects of drift it would be necessary to replicate such lines or ensure that the size of a single population was adequate.

The aim would be to develop divergent populations so that one could rapidly obtain a general yes or no answer as to whether a physiological trait would be likely to be useful, not to measure genetic associations precisely; (it is most unlikely that the genetic regressions would be the same in each replicate anyway). Without the need for elaborate recording of all traits every generation or to maintain unproductive lines such populations need not be expensive to run. Once established they would be available as a resource for research and to improve commercial populations by crossing if desired.

The feasibility of selecting sheep for high ovulation rate and changed seasonality is not in question. However, to be credible it is perhaps necessary to consider the two cattle traits in a little more detail.

Multiple ovulation in cattle: Selection for twinning by L.R. Piper and B.M. Bindon will test this approach. The basic problem is to establish a base population with a sufficiently high incidence of twin ovulations for subsequent progress to be feasible. The incidence of twinning in some sire groups might be 10%; with extreme selection among females and the use of such bulls, the incidence of twin ovulations in their daughters might be 10%. One then has the opportunity to select twin ovulating females as dams without the need for many laparoscopies.

Another approach would be to use C.H. Waddington's ideas of genetic assimilation. One could for example use sufficient antisteroid serum to raise the ovulation rate to an acceptable level to facilitate selection, then progressively decrease the dose until the trait was universally expressed.

High conception rates in cattle: This is probably the most adventurous but with conception rates of down to 50%, it is potentially very worthwhile. The aim might be to exploit differences among daughter half-sib groups to establish a base population and to reduce environmental variation by controlling aspects of timing. Remember Bradford did it in mice.

All of these selection programmes need cost very little as the animals will be of commercial value and can be kept by conventional husbandry. The long standing argument that the heritability of reproductive traits is so low that rates of change would not be high enough to be acceptable is itself unacceptable. It merely attempts to give credibility to inactivity.

The first component of the strategy is the establishment of appropriate populations for study. Once these are available, it will be possible to evaluate the biological importance of physiological variables, that is their potential as criteria for selection.

The second component would be the further study of both specific and comparative physiology. Such studies would be the basis of new criteria for evaluation in the resource populations and of possible biochemical criteria to screen for advantageous recessives.

To effect such a strategy it is perhaps relevant to ask why it has not been pursued in the past. Perhaps there has been too high an expectation that eventually enough knowledge would be available for the 'correct' solution to be described. The first step might then be the realisation that there is no such thing as the answer, current concepts of drift and knowledge of variation among replicates indicate that with manageable populations, correlated responses are unlikely to be identical, so that it is only possible to arrive at generalisations and expectations. Indeed such variation could be used and the population with the most favourable correlated response to indirect selection offered to commerce.

B. *To Summarise*

There is the present possibility of direct selection for the reproductive traits themselves. In the medium term there is now a reasonable possibility of using physiological criteria of polygenic merit and biochemical criteria for major genes including rare recessives which are neither age nor sex

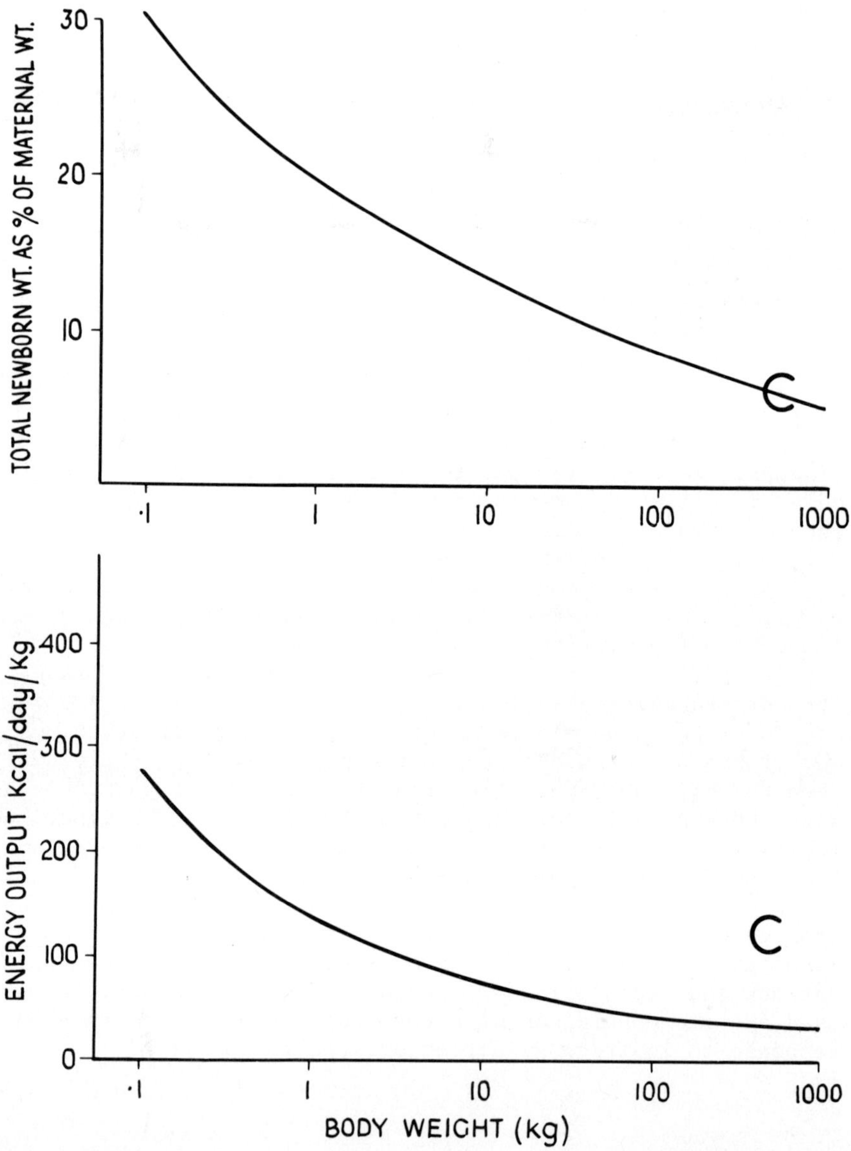

FIGURE 5. The decline in the relative reproductive and milk output as body weight increases in mammals; the average reproductive performance of cattle and their markedly superior milk yield (C).

limited. What next, or more specifically, are there prospects for direct manipulation of the genome? In the long term one has to say 'yes if' - if it is cheap enough to be used. A first step might be the transfer of individual genes across species - the transfer of the Booroola gene to cattle within the stated 25 years might be a reasonable target. One can readily see improved criteria being incorporated into conventional selection programmes during this period. The direct biochemical reassortment of the genome has to be both more cost effective than population selection and to beat the manipulation of microorganisms. Microbial milk might be just as likely as reassorted dairy cattle. In any event physiological knowledge of the expression of favourable alleles is an important step on the road to 'genetic engineering'.

C. A Final Challenge

"Further possibilities": What might be possible? The mention of dairy cattle sets up a challenge. Milk yield like conceptus weight decreases as a proportion of adult body weight as the body weight of a species increases. Selection of the dairy cow, however, has led to a daily energy output in milk per kg of body weight as high as that of animals only one hundredth of their size; putting it the other way round, the best cows produce four times as much energy per day as would be expected for animals of their size (Hanwell and Peaker, 1977). By contrast total conceptus weight is close to expectation (Fig. 5). Cattle have therefore to be selected to produce quads without any loss of individual birth weight before the genetic improvement of reproduction catches up with the genetic improvement of lactation. Are four teats a coincidence?

ACKNOWLEDGMENTS

We are pleased to acknowledge the contribution of discussion with many friends in various laboratories to the formation of the ideas presented. In addition to colleagues in ABRO, the University and the MRC at Edinburgh, we would particularly like to thank B.M. Bindon and L.R. Piper of CSIRO, Australia, G.E. Bradford, Davis, California and J. Thimonier and P. Mauleon of INRA, France.

REFERENCES

Berger, P.J., Shanks, R.D., Freeman, A.E., and Laben, R.C. (1981). *J. Dairy Sci. 64,* 114.

Bindon, B.M., and Piper, L.R. (1982). *Proc. Wld. Congr. Sheep and Beef Cattle Breeding, New Zealand.* (in press).

Bindon, B.M., Cummins, L.J., Piper, L.R., and O'Shea, T. (1981). *Proc. Aust. Soc. Reprod. Biol. 1981,* p. 79 (Abst.).

Bittman, E.L., and Karsch, F.J. (1981). *Biol. Reprod., Suppl. 24, 104A* (Abst.).

Bradford, G.E., Barkley, M.S., and Spearow, J.L. (1980). *In* "Selection Experiments in Laboratory and Domestic Animals" (A. Robertson, ed.), p. 161. Commonwealth Agricultural Bureaux, Farnham Royal.

Cahill, L.P., Saumande, J., Ravault, J.P., Blanc, M.R., Thimonier, J., Mariana, J.C., and Mauleon, P. (1981). *J. Reprod. Fert. 62,* 141.

Carr, W.R., and Land, R.B. (1982). *J. Reprod. Fert.* (in press).

Charlton, H.M., Parry, D., Halpin, D., and Webb, R. (1982). *J. Endocr.* (in press).

Dorrington, J., and Gore-Langton, R.E. (1981). *Nature, Lond. 290,* 600.

Eisen, E.J., and Johnson, B.H. (1982). *Genetics* (in press).

Forbes, J.M. (1975). *J. Endocr. 64,* 549.

Franchimont, P., and Channing, L.P. (eds.) (1981). "Intragonadal Regulation of Reproduciton." Academic Press, New York.

Goodman, R.L., Pickover, S.M., and Karsch, F.J. (1981). *Endocrinology 108,* 772.

Hanrahan, J.P., and Quirke, J.F. (1982). *Proc. Wld. Congr. Sheep and Beef Cattle Breeding, New Zealand.* (in press).

Hanwell, A., and Peaker, M. (1977). *Symp. Zool. Soc. Lond. 41,* 297.

Kacser, H., and Burns, J.A. (1979). *Biochem. Rev. 7,* 1149.

Kacser, H., and Burns, J.A. (1981). *Genetics 97,* 639.

Karsch, F.J., Goodman, R.L., and Legan, S.J. (1980). *J. Reprod. Fert. 58,* 521.

Kay, R.N.B. (1979). *Agricultural Research Council Research Review 5,* 13.

Land, R.B. (1981). *Livestock Prodn. Sci. 8,* 203.

Land, R.B. (1982). *Proc. Wld. Congr. Sheep and Beef Cattle Breeding, New Zealand.* (in press).

Land, R.B., and Carr, W.R. (1979). *In* "Genetic Variation in Hormone Systems" Vol. 1 (J.G.M. Shire, ed.), p. 89. CRC Press, Inc., Boca Raton.

Land, R.B., Carr, W. R., McNeilly, A.S., and Preece, R.D. (1980). *J. Reprod. Fert. 59*, 73.

Peters, R.R., Chopin, L.T., Leining, K.B., and Tucker, H.A. (1978). *Science 199*, 911.

Piper, L.R., and Bindon, B.M. (1982). *Proc. Wld. Congr. Sheep and Beef Cattle Breeding, New Zealand.* (in press).

Purser, A.F. (1973). *A.R.C. Animal Breeding Research Organisation Report*, p. 24.

Ryan, K.D., and Foster, D.L. (1980). *Fedn Proc. Fedn Am. Socs exp. Biol. 39*, 2372.

Walkley, J.R.W., and Smith, C. (1980). *J. Reprod. Fert. 59*, 83.

Webb, R., and England, B.G. (1982). *Endocrinology* (in press).

Wilmut, I., and Sales, D.I. (1981). *J. Reprod. Fert. 61*, 179.

Wolfe, H.G., Bartke, A., Amador, A., van Sickle, M., Dalterio, S., and Brown, D. (1981). *J. Endocr. 90*, 367.

Womack, J.E., and Lu Shun Yan, D. (1981). *Science 212*, 63.

The Significance of Reproductive Biology to the Genetic Improvement of Farm Animals

David Lindsay

Department of Animal Science and Production
University of Western Australia
Nedlands, Western Australia

Progress in the genetic improvement of genetic traits in domestic livestock is hampered by poorly defined objectives. Reproduction is a multistep process; progress in one step may be correlated negatively with progress in another especially where environmental or managemental levels are unfavourable. There is merit in breeding for lack of anoestrus, seasonal or lactational, to increase productivity per annum. Mothering ability becomes increasingly important as fecundity rises but selection criteria have not yet been defined. Ovulation rate can now be determined rapidly by laparoscopy and its heritability is reasonably high. More needs to be known of its final effect on numbers of offspring which are weaned. Selection for multiple births has great potential but many prolific breeds have too wide a range of birth types and poor survival in litters larger than twins. Breeds of sheep and goats exist in which the litter size is predominantly twins but they have not yet been exploited. The ideal reproducing sheep or goat would be one with a short non-pregnant period, a multiple litter limited to twins and good mothering ability. The physiological bases for these are imperfectly known and are urgently needed for efficient genetic progress.

FUTURE DEVELOPMENTS IN THE
GENETIC IMPROVEMENT OF ANIMALS
ISBN 0 12 078830 6

I Introduction

The close relationship between reproductive biology and genetic improvement in farm animals has been recognised for a long time. A high reproductive performance enables a reduction in generation length and/or an increase in selection differential. But this advantage has been only minor compared with the impact of certain reproductive techniques, particularly artificial insemination, on the practical development of genetic programmes. The perfection of artificial insemination in cattle was the key which allowed the vast and spectacular genetic programmes in dairy cattle throughout the world to open up. It enabled genetics to be demonstrated successfully to farmers who formerly had little interest or faith in the discipline. The progress of genetics in other branches of livestock is similarly dependent on reproductive techniques which allow not simply widespread use of individual animals, but access to flocks and herds whose owners would not otherwise accept modern genetic principles.

A more recent trend has been a reversal of this dependence with an increase in attempts to improve reproductive performance of animals using genetics.

In primarily primiparous species pessimism surrounded this approach for many years, associated with a low heritability of multiple births. This has been dissipated to some extent by the 'discovery' of certain strains which are already highly fecund and therefore make cross-breeding attractive as compared with the former, slow approach of selection within poorly performing breeds. In the Australian Merino in particular, a further boost has come from recognition that what was hitherto exclusively a wool breed, has suddenly become an excellent animal for the Middle East meat trade. Good reproduction has been recognised as essential for a profitable enterprise in surplus animals.

Superficially, the genetics-reproduction interaction is simple. Higher reproductive rates mean higher selection pressures and quick genetic gain. In fact, as we find ourselves at what is probably only the beginning of the exploitation of the genetics of reproduction, a number of problems in definition of objectives are beginning to emerge which will complicate such exploitation.

It is appropriate at this meeting to consider some of the implications of selection programmes for reproduction as a basis for future breeding policy.

II Indirect Responses in Reproduction to Selection for Productivity

The classic example in turkeys, in which selection for body size developed birds too big and cumbersome for breeding, illustrates dramatically that reproduction is by no means independent of other productive traits. In the larger animals, new and more subtle differences are showing up. For example, P. le Neindre (personal communication) in France has demonstrated that a breed selected principally for milk production - the Friesian - has also been selected against good maternal behaviour. He compared a hardy mountain breed - Salers - with Friesians, and found that one component of the superior milk production of the Freisian is its ability to adjust to the withdrawal of its calf. In fact, cows which show strong maternal instinct and concern themselves about their nursing calves, are bad producers of milk and are likely to be culled. By contrast, the hardy breed which is well adapted to rearing its own calf cannot be milked unless in the presence of its calf. The calves themselves also differ. Friesian calves have a relatively poor sucking reflex and can be trained, after a few attempts, to drink from a bucket. The Salers calves may take months to train, and they retain a strong instinct to seek their mothers. High milk production is a strong, positive characteristic in reproductive fitness, especially where multiple births are anticipated. If, in a cross-breeding programme for extensively reared beef animals, this is achieved at the cost of reduced maternal behaviour of the mother and lower sucking drive in the calf, then little will have been achieved.

III Direct Selection for Reproduction

The simple measurement of the number of offspring after a single gestation is seldom a good criterion of reproductive efficiency. It is therefore probably not a good criterion for selection for reproduction. In the pig industry, it has been recognised for some time that the number of pigs per litter is far less important than the number of pigs reared per year. One is a component of the other, but many factors intervene to modify the annual productivity. It can be shown, for example, that in a sow of average litter size, a modest reduction in the period of lactation from eight to five weeks can raise the annual productivity more than can one extra pig per litter. In the case of the pig, the simple reduction in duration of lactation is far easier to achieve than a genetic gain of one pig per litter.

Let us examine the principal components of annual reproductive performance, and look at prospects for genetic improvement.

A. *Ability to Conceive at Any Time of the Year*

Long periods of anoestrus restrict the potential of many breeds of sheep and goats and, to a lesser extent, cattle. There is some variability within breeds and a great deal of variability among breeds. Hafez (1952) proposed that the length of anoestrus in breeds of sheep was related to the distance of the origin of the breeds from the equator. Breeding activity and inactivity in sheep are certainly controlled by changes in day length, so it is assumed that breeds developed near the equator have more or less continuous activity. Hafez's theory has many exceptions which suggests that both within and between breeds there is scope for selection. For example, the Romanov breed which originated at higher latitudes than many other European breeds, has about the same length of breeding season as they do but it starts and ends much later. The Merino is often quoted in northern hemisphere texts as a breed reputed to breed all year round. In fact, it has a 'spontaneous' breeding season little different from that of many other breeds but can be induced out of anoestrus by the introduction of rams. This highly important behavioural variable has been termed the 'male effect' - and has been shown with other breeds of sheep and with tropical goats (Chemineau, 1982).

The length of the breeding season depends on whether the male effect is effective or not. Females which remain in the permanent presence of males do not show the effect and remain in anoestrus for long periods. Females previously isolated respond to males almost immediately and begin to ovulate within a day or two of the introduction of the male.

A further complicating factor is that the male effect has also been reported to raise the ovulation rate at the first ovulation. Whatever the complicating factors, the 'male effect' in sheep and goats is a powerful and almost costless means of improving the potential for annual productivity in some breeds.

We are still trying to understand the physiology of the phenomenon, but it is clear that certain breeds are more amenable to its use than others, and certain animals within breeds are more responsive than others. There is thus a genetic dimension which requires systematic study. Eventually, breeding programmes which incorporate the ability of females to respond to the male may offer a new and simple means of improving the potential annual productivity of sheep and goats which is at present confined to the narrow limits of an annual breeding season.

Apart from seasonal anoestrus, the non-reproductive phase can be lengthened by lactation. The hormonal milieu of lactation is not particularly conducive to optimum chances of conception. This is most apparent in the sow, where anoestrus is absolute during lactation and fertility does not recommence until after weaning. It is believed that this is also true for some breeds of sheep and goats. It is certainly not true for all breeds of sheep and goats, as Obst *et al.* (1980) have shown for some Indonesian breeds.

In sheep in particular, breeding and selection against long periods of seasonal or lactational anoestrus offer potentially powerful means of increasing the reproductive capacity of flocks. Provided that a satisfactory feed plan can be arranged, a ewe rearing one or two lambs at a time but more than once a year, has a much greater chance of success than a ewe attempting to rear three at a time, once a year. For example, the Indonesian sheep studied by Obst *et al.* (1980) and by A. Winantea and J. Juliani (personal communication), for which distributions of multiple births are shown in Figures 1 and 2, show little or no seasonal or lactational anoestrus. In both cases, with an average lambing rate around 2 and 1.6-1.8 lambings per year the ewes are more productive on an annual basis than those breeds with significantly higher multiple birth rates.

Information has accumulated, particularly in cattle, that the heritability of the interval - parturition to conception - is low. This probably holds also in sheep and goats but, as with ovulation rate, certain breeds exist which are vastly different from others. It is therefore reasonable to assume that cross-breeding for short lactational anoestrus may be successful. Certainly, the potential advantages warrant some genetical research into this largely neglected area. Such research should be conducted in conjunction with a study of the genetics of response to the male effect. The ultimate goal of such studies would be the development of strains capable, not only of breeding at any time of the year regardless of their lactational state, but also capable of being synchronised. Such synchronisation would, in turn, open the way to the use of management systems designed to improve survival of lambs.

B. Mothering Ability

In the absence of elaborate artificial feeding techniques the final production of lambs from multiple-lambing flocks depends on the ability of the mothers to recognise, accept, feed, and rear the offspring. There is little evidence that this mothering ability is positively correlated with multiple births and will therefore need to be given much more careful attention in its own right. We have already seen, in fact, that selection for dairy milk production in Friesian cows leads to selection against strong maternal behaviour. Most of the prolific breeds come from regions in which flock sizes are small and intensification is high. Maternal care by the mother herself is therefore less of a consideration than in an extensive system of management. If lambing percentages can be extended to extremely high levels and this is coupled with synchronisation of lambing, the economics of the total supervision and artificial rearing may be acceptable. Otherwise, and perhaps in addition, a system of selection which gives heavy emphasis to mothering ability is essential. Schemes such as the New Zealand High Fertility Romney project, in which the selection criterion is the ability to *rear* twin lambs, obviously have a strong maternal component. But there have been no attempts so far to study what components of maternal behaviour are heritable and what can be ameliorated by simple adjustments to management. This information is essential if we are to develop efficient breeding schemes which include the ability of ewes to rear lambs. Certainly, such research must go hand in hand with improvement of the animal's ability to produce offspring in multiples.

C. Ovulation Rate

The recent introduction of the laparoscope has opened up a new possibility for a selection criterion associated with reproduction. It has the dual advantage of allowing examination of the reproductive potential of a female soon after puberty and at the same time discovering her 'upper limit' before uterine and environmental factors reduce the eventual productivity. This technique is too new to have been extensively studied but a number of versions of schemes using laparoscopy have been initiated in the last few years. It is now known that the relationship between ovulation rate and litter size following the same ovulation is not as high as one might expect. In pigs, this correlation is only 0.48 (Paterson *et al.*, 1980). Fertilization failure and early embryo survival are thus highly variable and unpredictable components in the sequence of events leading to reproductive success. They undoubtedly account for much of the low heritability of multiple births. Even though ovulation rate apparently has a reasonable heritability and we can make progress in ovulation rate by selecting for it, we still have to find out whether, and under what management conditions, this can be translated into improved prolificacy. Reproductive biologists have been challenged anew to look at the physiology of embryo loss and survival, and of offspring loss and survival, as this is the key to the success or otherwise of schemes which use ovulation rate as a selection criterion. It is likely that ovulation rate may have to be coupled with some other criterion associated with survival of embryos and offspring to achieve maximum gains through this technique.

D. Litter Size

In recent years the interest in improving the prolificacy of sheep and pigs has centred almost exclusively on increasing the number of offspring per animal per parturition. Selection programmes based on within-breed selection have shown some worthwhile gains but have confirmed old predictions that heritability is low. The most spectacular improvements have come from the identification of strains and breeds with natural high fecundity. The most notable of these in sheep are the Booroola, the Finnish Landrace and the Romanov, all of which are capable of producing litters of three or more. In the pig, some Chinese breeds, the Mei Shan and the Jai Xing, regularly have litters of 20 or more. The incorporation of these in cross-breeding programmes has been advocated and practised for a number of years on an experimental basis and the fecundity of the female offspring has certainly increased dramatically over that of their non-prolific mothers. Research is under way in a number of centres to find ways of identifying the presence of genes for prolificacy in male animals in order to increase the overall efficiency of selection.

It is appropriate at this point to raise two questions;

(1) Are we selecting for multiple births in the best way?

(2) Is selection for multiple births the best way of increasing prolificacy anyway?

1. Are We Selecting for Multiple Births in the Best Way? For many years, Australian sheep farmers have quoted the adage that 'twins are bad news' and it is preferable to 'hit one on the head' to allow the mother to rear one lamb satisfactorily. I have never had one admit that he actually carried out this threat, and a well fed ewe should and can rear twins reasonably successfully. Nevertheless, I take up the Australian sheep farmers' stance when it comes to more than twins. For a number of reasons, losses soar astronomically, even in well-nourished flocks, when lambs are born as triplets or quadruplets. Systems for cross-fostering, or more usually artificial rearing, have been developed in an attempt to counter this in the prolific breeds but the economics of such systems are, at best, marginal. Despite this, the most common method of applying selection pressure for multiple births is to select those animals which were born as triplets, quadruplets, or even quintuplets - or, more correctly, the survivors of such multiple births. If progress is being made in selection for multiple births the problems of rearing are confounded with each successive generation.

Let us look at other options. The traditional Australian sheep farmer wants 100% of ewes to have one lamb each. There seems to be a negative correlation between the ovulation rate and the proportion of ewes in a flock which are dry. Thus it seems that the price of getting a low number of dry ewes is to have many ewes which have twin ovulations and so some, at least, which have twin lambs. If I were an Australian farmer I would like to see 100% of

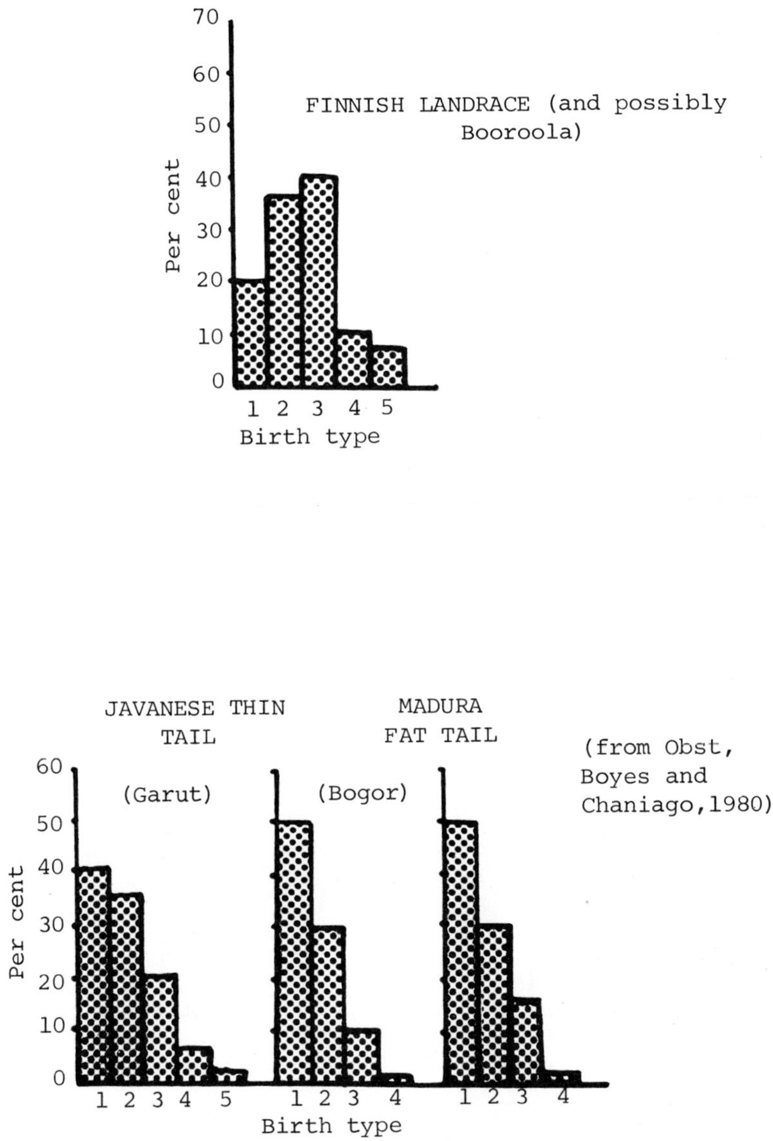

FIGURE 1. Distribution of multiple births in fecund breeds of sheep.

ewes lambing and each having twins. If I were to achieve the impossible goal of 100% of births being twins, results of a number of comparative studies suggest that I would also have very few dry ewes. But how impossible is the goal of almost all births being twins? If we are dealing with the presently popular prolific breeds, there would be no chance. Figure 1 shows the typical distribution of birth types for Finnish Landrace, Romanov and Booroola breeds. In each case the distribution is wide and the predominant birth type is triplets. Perhaps our sights are too high. Figure 1 also presents the distributions of birth types for an Indonesian breed of sheep studied by Obst *et al.* (1980). In this case the average birth type is twin, and the predominant birth type is a single. Unfortunately, the distribution is still too wide and there are too many multiple birth types greater than twins. Two slightly less prolific breeds, the Barbados Blackbelly, studied by Devendra (1977), and the East Java Fat Tail, studied by A. Winantea and J. Juliani (personal communication) (Fig. 2), present a third option which I believe to be the most promising yet.

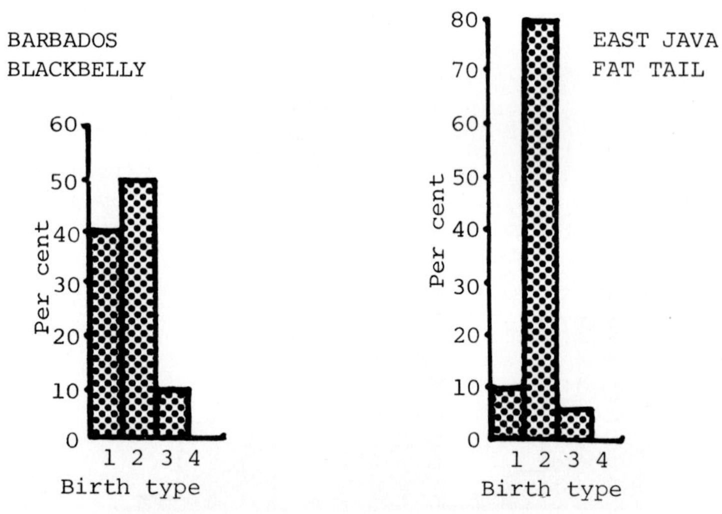

(from Devendra, 1977)

(from Winantea and Juliani, personal communication)

FIGURE 2. Distribution of multiple births in fecund breeds of sheep.

These are the only breeds so far studied in which the predominant birth type is twins - in one case 83% - and in which the distribution is very narrow: neither the genetics nor the physiology of multiple births in the two breeds has been studied. Their effectiveness for crossbreeding is therefore unknown but for the moment they, or other breeds which may meet the same criteria, rest as potentially the most practical of the prolific breeds so far studied.

2. Is Selection for Multiple Births the Best Way of Increasing Fecundity? From the foregoing it is apparent that prolificacy has many components. The most spectacular is undoubtedly multiple births. This may be the reason why other components have been largely ignored. The ideal calving interval is considered to be 365 days and a cow complying with this ideal would be pregnant 3/4 of her reproductive life. Sheep or goats are normally expected to produce offspring only once a year. Thus they are pregnant only five-twelfths of their reproductive life. We have a large, virtually untapped source of reproductive potential lying in the non-reproductive phase. Breeds exist which have the potential to form the basis of new strains capable of increasing prolificacy by accelerating the reproductive turnover in these species. The principal argument against this approach is that a single parturition per year can be timed to take advantage of the most favourable seasonal conditions. More frequent parturitions would lead to neonatal mortalities caused by unfavourable nutrition and climate. It should be remembered that when we talk about high fecundity we are not talking about animals running under marginal conditions of survival. This applies equally to the development of multiple ovulating animals. The advantage of high fecundity through short intervals between parturitions is that the extra costs of out-of-season management can often be offset by high out-of-season prices for the resulting offspring.

For a given environment and level of management, the reproductive output of a group of adapted animals appears to be remarkably steady. In fact a given level of reproduction is probably part of the overall adaptation of the population to the environment. If some of the components of the reproductive process are varied, others may shift to maintain the steady state. For example, an increase in ovulation rate through genetic, nutritional, or hormonal means may well be nullified by an increase in embryonic or post-natal death, resulting in little improvement in the number of offspring produced. This has been the average result of a host of experiments in pigs, sheep and cattle where ovulation rate has been increased artificially. Why should it be remarkably different if we increase ovulation rate by genetic means? Clearly, management of the animals must go hand in hand with genetic improvement in the reproductive inputs if the benefits of such genetic programmes are to be felt. In other words, no thought of improving any of the reproductive inputs of herds or flocks by genetic means should be attempted without equally careful consideration being given to changes, perhaps massive changes, in the environment and management to allow phenotypic expression of genetic gains. The costs and benefits of one should not be counted without those of the other.

IV The Future of Breeding for Increased Reproduction

Reproduction is the complex interaction of a number of physiological processes. There is not yet a breeding programme which had been designed to take into account all of the physiological processes. Some of these processes have low heritabilities so that progress is slow in any one of them. A multi-trait selection programme would obviously lead to even slower progress in each trait. We can regard present schemes based on selection for large litters as being useful for cross-breeding. Maybe if management and supervision reach levels which it is presently difficult to imagine they will ever attain, these highly fecund animals may even be the types of animals which we seek. Meanwhile, the more modest ideals of schemes in which we seek to select on ability to rear twin lambs may be more realistic in extensive systems. However, selection schemes will make slow annual progress, first because of the large environmental component of their variability and, secondly, because the basic stock lack desirable features of other breeds which would enhance productivity. For example, breeding for lack of anoestrus in Romney Marsh sheep is difficult because there are not enough animals which breed all year to select out a nucleus flock. Thus cross-breeding to introduce certain characters appears desirable if an animal incorporating most or all of the favoured components is to be developed reasonably rapidly. The problem is that in most cases the biology and particularly the physiology of the desirable components which we have been discussing is imperfectly known. For this reason, the success of selection for them is likely to be problematical at this stage. Never has there been a more appropriate time for a cooperative effort between geneticists and reproductive biologists. Without it, I can't see a great deal of progress being made in the reproductive performance of our domestic animals.

REFERENCES

Chemineau, P. (1982). *J. Reprod. Fert.* (in press).
Devendra, C. (1977). *World Rev. Anim. Prod. 13,* 31.
Hafez, E.S.E. (1952). *J. agric. Sci., Camb. 42,* 189.
Obst, J.M., Boyes, T., and Chaniago, T. (1980). *Proc. Aust. Soc. Anim. Prod. 13,* 321.
Paterson, A.M., Barker, I., and Lindsay, D.R. (1980). *Aust. vet. J. 56,* 547.

Discussion Summary

The major issues considered related to the lack of genetic information on many aspects of reproduction, and particularly on the genetic interrelationships among them, and with production traits.

In some environments, particularly under extensive management, a high twinning rate in sheep is disadvantageous because both lambs commonly will not survive. If average reproduction rates were to be increased, the ability to control litter size according to level of nutrition would be an advantage in those areas where availability of pasture was unpredictable. Any attempt to do this, however, presumes knowledge of the genetic control of litter size and maternal ability. The results noted for the East Java Fat-Tail sheep of Indonesia, with a very high percentage of ewes producing twins, suggest a unique genotype and physiology that warrants intensive study. The genetics of the maternal contribution to neo-natal survival is largely unknown. However, data from Merinos in Australia (differences between half-sib ewe groups) suggests a moderate heritability (about 10%), and studies are in progress to analyse the factors that trigger ewe maternal behaviour essential to lamb survival.

Limitations to the genetic improvement of reproductive rate may be imposed by combined effects of embryonic and neonatal losses. In pigs, uterine capacity may be a limiting factor, and there has been little or no effective selection for litter size in pig improvement programs. Yet genotypes are known (breeds of Chinese pigs) that have very large litters. It appears that no information is available on ovulation rate or embryo wastage in these breeds, and study of their reproductive physiology should be both exciting and valuable.

Of course, it is recognized that these breeds, unique for specific reproductive traits, are or may be inferior for other traits or that there may be undesirable genetic correlations between these reproductive traits and other economically important traits. Integrated study of their reproductive physiology and genetic basis is vital. Further, these sheep and pig breeds are so extreme, relative to breeds that have been investigated more intensively, that the differences may be due to one or only a few gene loci. Could the difference in each case be in just one regulatory locus?

Insufficient data are available on genetic correlations among components of reproduction and between each of these and productive traits to allow definition of optimum selection criteria for the improvement of reproductive performance. For example, although considerable attention has been devoted to testis size, it was not certain what effect selection for testis size had on age at puberty in males and females. Further, selection on testis size in males at different ages may have different effects. The primary motivation for this work was to explore relationships between male and female fertility. This remains a major imperative and it seems likely that the optimum approach to this problem would be to create divergent lines selected for various components of male or female reproduction and then to study correlated responses at the biochemical and physiological level. Such lines would allow study of additional components as they are recognized or as new techniques become available.

Finally, genetic changes to reproductive traits that could be of economic importance, but which have received scant attention, include prolonged gestation in pigs, to improve piglet survival, and shorter gestation length in cattle, to increase production per day of life and in some breeds, to decrease calf losses due to dystocia.

Part 5
Economic Aspects of Developing Breeding Objectives

Economic Aspects of Developing Breeding Objectives: General Considerations

J. W. James

School of Wool and Pastoral Sciences
University of New South Wales
Kensington,
New South Wales

A number of problems in defining breeding objectives are discussed. It is suggested that the difference between costs and returns will usually be preferable to their ratio. A recently proposed economic model is considered to be an inappropriate framework for estimating economic weights. It is stressed that breeding objectives and selection criteria must be clearly distinguished if logical procedures are to be followed. The discounted cash flow method is a useful way of allowing for time differences in realisation of returns, but does not seem always to be easy to apply. The detail which should be used in specification of objectives is discussed, and it is tentatively suggested that the level of analysis should be that of items involving income and expenditure.

I Introduction

The most important decision of all in any animal improvement program is the choice of the breeding objective. If the objective is badly chosen, then any genetic progress which is made will be in the wrong direction, and may even lead to economic deterioration of the population. For example, those

FUTURE DEVELOPMENTS IN THE
GENETIC IMPROVEMENT OF ANIMALS
ISBN 0 12 078830 6

breeders who did not read the signs correctly in the early part of this century and sought to develop further the "lard type" pig may have succeeded in their objective, but were not successful breeders from the viewpoint of the pig industry.

Until fairly recently, scientific animal breeders have not given a great deal of *systematic* attention to the choice of breeding objectives, perhaps because this problem is essentially a non-genetic one, and geneticists have been in the majority of scientists involved in animal breeding. Furthermore, they were understandably concerned to develop and test methods of genetic improvement which could be applied in general. Nevertheless, one of the major causes of conflict between traditional and scientific breeders has been the definition of objectives. Scientific breeders have questioned the pursuit of aims which conferred no clear economic benefit. There have also been disputes over selection criteria, and the distinction between objectives and criteria has not always been made as clearly as it should be. To avoid any misunderstanding, I shall define how I use these terms. The objective is what the breeder seeks to maximise in his population. The traditional breeder may seek to maximise resemblance to a conceptual ideal animal, but more often the objective will be some measure of economic performance. In this paper I shall assume that the second form of objective is to be considered. Selection criteria are traits of the animal or its relatives which are actually used by the breeder in deciding whether or not it will be used as a parent of the next generation. In principle, these criteria will be chosen, subject to constraints of cost and time, as those which will give most progress in the desired direction. In a sense, the breeding objective has to be defined before we can make a rational choice of selection criteria.

In practice the objective is often not formally defined, and selection criteria are often chosen as being economically important traits, some informal consideration of their relative importance being used to decide on relative emphasis to be given to different characters. However, if no consideration is given to possibly important economic characters because they are not to be used as selection criteria, the consequences can be serious if the "forgotten" trait happens to have disadvantageous genetic correlations with the selection criteria. Therefore, I believe the breeder should make a serious effort to define his objective in a formal manner. Land (1981) has emphasised the problems of forecasting future requirements, and has proposed an "alternative philosophy" of developing strains with divergent biological characteristics. Even then it would, I think, be necessary to decide what were important traits, not only in present, but in future conditions. It is not clear to me that this proposal, interesting as it is, will overcome our difficulties.

II Some Basic Problems

Moav (1973) drew attention to a rather difficult problem in evaluating breeding programs, namely, who benefits? He argued that the community as a whole was likely to be the major beneficiary of industry wide improvement. If there is a fixed market for milk, increased productivity is likely to lead to fewer dairy cattle and thus to fewer dairy farmers. The surviving dairy farmers will be better off, but those forced out of the industry may not be. These considerations are important in deciding who should pay for national programs, but do not seem to me to have much effect on the determination of relative economic values.

In all industries, competition between producers ensures that the more efficient will do better. The fact that the industry as a whole may not gain financially from genetic improvement can then not be used as an argument for not seeking such gains. For, if this is not done on an industry-wide basis, there is an opportunity for those who do make improvement to obtain a larger share of the limited market. Dickerson (1976) has argued that a breeder should seek to minimise cost per unit product, essentially on the ground that, as productivity rises, prices will fall. He essentially assumed a fixed market since he used the unit of product as the basis. In practice, this is not really accurate, since there is never really only one product, and one must weight the different products by their relative values. The assumption that markets are fixed at present levels for all products is somewhat strong. Dickerson suggested the use of the ratio COSTS/RETURNS as breeding objective. Harris (1970) had discussed possible use of this ratio, of the inverse ratio RETURNS/COSTS, and of the gross margin RETURNS-COSTS. Choice of the COSTS/RETURNS ratio is recommended on the basis that the market for products functions as the limitation, so that total returns cannot be increased, so one should minimise costs. However, we have in the recent past a good example of a large change in returns in the growth of the broiler industry. Here, rapid genetic change (and rapid change in husbandry) led to a major expansion in the market because prices fell and total returns increased. In the larger species such rapid progress and accompanying changes are unlikely, but the principle is still valid that RETURNS can not be considered as a constant. The reciprocal ratio RETURNS/COSTS suggests that expenditure by the producer is the limiting factor, and that his objective is to get as high a return for this expenditure as possible. For extensive grazing properties this may be nearly true, the provision of feed, primarily through capital expenditure for land purchase, and perhaps also pasture improvement or fodder cropping, being the main limitation. If the input is fixed, the ratio of output to fixed input is an obvious objective. However, it should be noted that neither of these ratios used as an objective will necessarily satisfy the apparent aim. Thus, for instance, the most effective way to decrease cost per unit product may be to increase output, rather than to reduce input. It would be necessary to impose a restriction on either numerator or denominator of the ratio in order to ensure that the objective did what was desired. But if (say) RETURNS is taken as fixed, then the only avenue for improvement is reduction of COSTS, and this will be achieved by using any of the three possible criteria discussed by Harris. Morris (1981) has pointed out that gross margin is probably the major incentive for change among Australian graziers, and my own preference is for the use of margin rather than either ratio.

Many agricultural products have their prices determined by supply and demand. Many producers prefer a reduction in costs to an increase in production on the grounds that, at least in the short term, a reduction in costs will not alter prices and will then lead to greater profit even with the same production. On the other hand, increased production is expected to lead immediately to reduced prices, and consequently does not guarantee increased profitability. But as pointed out above, these concerns do not lead to a decision on how to combine COSTS and RETURNS, since no particular form is certain to concentrate on reducing inputs rather than increasing outputs. Ideally, we would like to have a full economic model which would allow us to predict changes in prices, costs, *etc.*, following from any genetic change in the population, and we could then determine the consequences for the incomes of producers and breeders. This is unlikely to be available in the near future, perhaps especially in those industries where political decisions rather than market forces are of major importance. Nevertheless, some approximation is necessary if objectives are to be defined.

The choice of how COSTS and RETURNS are to be combined does affect the relative economic weights given to different traits. This is easily seen by considering an artificially simple example. Suppose there is a single output W and a single input F whose prices are a and b respectively. Then RETURNS is given by

$$R = aW$$

and COSTS is given by

$$C = bF.$$

Then the partial derivatives of R-C are given by

$$\frac{d}{dW}(R - C) = a$$

$$\frac{d}{dF}(R - C) = -b$$

and the weight for F relative to that for W is -b/a. On the other hand, the partial derivatives of C/R are given by

$$\frac{d}{dW}(C/R) = -b\bar{F}/a\bar{W}^2$$

$$\frac{d}{dF}(C/R) = b/a\bar{W}$$

so that the relative weight for F is $-\bar{W}/\bar{F}$ where $\bar{W}$ and $\bar{F}$ are the means. Thus for R-C the weights depend on prices but not means, while for C/R the weights depend on means but not prices. There is thus a very large and important difference between them. If there are in addition overhead costs k in this model, it is easily shown that this does not alter the relative weights for R-C, but that for C/R the relative weight for F becomes $-\bar{W}/(\bar{F} + k/b)$. Thus the ratio is senstive to the relative importance of fixed and variable costs.

This is only a simple and artificial example, but does illustrate the importance of the decision as to how COSTS and RETURNS are to be combined. It is the difficulty of making such decisions which often leads to the suggestion of a "biological" objective such as lean tissue feed conversion in pigs (Fowler, Bichard and Pease, 1976). However, such an objective is biological only in the sense that it consists of a very simple economic model, with feed being the sole cost and lean tissue the sole product. It is therefore a simplified form of the objective proposed by Dickerson (1970, 1976). While it may be conceptually simpler and yet closely related to a more fully defined objective, I do not see that it has real advantages over Dickerson's form. I must add that I feel uncomfortable with an objective which does not depend on relative prices of inputs and outputs.

In fact, no industry is monolithic and with competition between producers the margin between COSTS and RETURNS will determine what producers seek in the animals they buy from breeders. Since only the stock used by producers will have an important influence on the whole population, I believe that what the breeder should aim to improve is the difference, RETURNS - COSTS. But it is well recognised that different producers have different aims, and breeders may well have different objectives to cater for the different requirements of various producers. For instance, some producers are risk avoiders, and might prefer to maximise the minimum margin rather than the expected margin. A sheep farmer who was a risk avoider might put less weight on number of lambs weaned because in bad years the high mortality among the increased number of twin lambs would reduce the margin to a very low value, even if the average over all years favoured high twinning rates.

III An Economic Model

Recently, Melton, Heady and Willham (1979) proposed a procedure for estimating economic values. This method has been discussed also by Thompson (1980) and by Goddard (personal communication). Basically, Melton *et al.* propose to maximise profit per animal (though other bases could be chosen) where profit per animal is expressed as a function of all inputs. The special feature of their model is that inputs fall into two classes, those supplied by the producer and those embodied in the animal. They assume that the producer is a profit maximiser and thus chooses both the producer supplied inputs and the animal traits in order to maximise profit. There is thus an implied price for each unit of each animal trait. It is of course not open to vary these animal traits as inputs freely, since they come as a complete package in the animal. Very likely, the particular combination of traits that the producer would prefer is simply not available. However, there is a purpose in the fiction.

If the levels of the traits have been chosen to maximise profit so defined, then by differentiating with respect to all the traits, we find a series of partial derivatives which can be equated to zero. These can be used to provide a series of equations involving the values of the traits and their means. It is assumed that the producers have chosen the means to satisfy the equations, knowing the trait values. However, we can use the equations to find the trait values, knowing the means. An alternative approach would be to express profit as a function of producer supplied inputs and animal traits, but not to include animal traits as inputs with implied prices. If the producer supplied variables are chosen so as to maximise profit under this model, then it is easily shown that the partial derivatives of profit with respect to the animal traits are exactly the same as the implied prices in the model of Melton *et al*. The whole purpose of the Melton *et al*. procedure seems to be to arrive at this result. In the simpler model, use of these economic weights leads to a maximum change in profit. In their model, however, use of these weights leads to an increase in the input costs, which hardly seems desirable. As Goddard (personal communication) commented, if profit is already maximised, why select for different trait values? According to Melton *et al*. the economic value of a trait depends on all aspects of the production process and prices and the key is the simultaneous and collective solution of the equations obtained by differentiating profit (as defined by them) with respect to producer supplied inputs and animal traits. But this gives exactly the same results as a simpler model in which the animal traits are not treated as inputs with implied prices, and the aim is to maximise change in profit. Since this simpler model gives a logical reason for using the resulting weights, and is not artificial, it seems entirely preferable. Thompson (1980) in discussing the model of Melton *et al*. pointed out that stationary points found by equating derivatives to zero are not necessarily maxima, and that their numerical example was such a case. However, this does not bear on the present discussion.

IV Separation of Objective and Criteria

I have previously (James 1981) emphasised the importance of distinguishing the objective from selection criteria and of making sure that every trait of economic importance is included in the objective. There have been many studies in which the distinction has not been maintained. Morris (1981) has used such an approach, while admitting its limitations, in finding an economic weight for carcase production in beef cattle. He considered various reductions in the economic weight to allow for possible correlated responses in feed requirement. This approach is legitimate only if there is a perfect genetic correlation between carcase production and feed consumption. In that case they are simply different aspects of the one trait. But if the correlation is less than perfect, this approach may lead to absurdities. For instance, suppose we had reduced the economic weight given to carcase production to allow for an assumed correlated change in feed consumption. This might work fairly well if selection were simply for yearling weight. But at least in principle it would be possible to devise a selection index which would increase carcase production while preventing any change in feed consumption. This would reduce the amount of genetic gain in carcase production. However, using the reduced economic weight we would have an inbuilt assumption that feed consumption had risen also. In fact, the pattern of correlated responses depends on the selection criteria which are used. We cannot therefore incorporate correlated responses into the breeding objective. The solution is of course simple. Just add feed consumption to the traits in the objective. I think the reason that some people are reluctant to do this is that feed consumption cannot often be used in practice as a selection criterion. But it is important to keep clear the difference between objectives and criteria, and it might even help to make this distinction more obvious if feed consumption were explicitly included in the objective.

A related approach is to estimate economic weights as done by Andrus and McGilliard (1975). Their method was to take data from dairy cows and calculate for each the profit per year of herd life. They then regressed this profit per year of herd life on various traits of the animals. The results are phenotypic regressions of profit on various traits. As is well known, the genetic and phenotypic associations between traits may well be different, though they are often similar. What their method does is to give an index to select for lifetime performance (Young and Tallis, 1961). What may have been a better approach would have been to calculate the regression of profit on parental phenotypic values. This would give a genetic selection index since it is, statistically, the best predictor of progeny profitability from parental phenotype. It would be even less accurate to attempt to estimate economic weights by regressions using different farms. Indeed, there will always be difficulties with regression approaches, since the predictor variables chosen for inclusion will be limited to those for which data are available, and may not include one or more important traits. Regression equations are more in the nature of selection criteria than objectives.

V Discounting of Returns and Costs

Up to this point I have avoided the question of how costs and returns over time are to be accumulated. The most common procedure has been to take all costs and all returns within a single year and use these. Ponzoni and Walkley (1981) calculated returns per ewe lifetime, but this is effectively the same as annual returns multiplied by the average number of years per ewe lifetime. By contrast, McClintock and Cunningham (1974) introduced the "discounted gene flow" method for giving relative economic weights to beef and dairy traits in dual purpose cattle. They used the purchase of a dose of semen as the basis on which to calculate resulting cash flows. For a ram buyer, essentially the same approach could be used. What one does is to calculate the pattern of cash flows and discount them all back to present value. From the viewpoint of a semen or sire purchaser this is eminently reasonable, since some returns will accrue earlier than others and thus be more valuable in relation to the buyer's investment. In the situation dealt with by James (1978), where both genetic gains and gains through culling in the present generation were considered, it is difficult to see how these two components of response to selection could be meaningfully combined without discounting. On the other hand, if one is concerned to devise an optimum sire buying policy, discounting does raise some problems. If the objective is non-linear, the stream of cash flows from a single selection is dependent on future selection decisions. This may not be serious in deciding on an objective at a single point in time, but in devising long term policies it may be important. James (1980) returned to the annual costs and returns basis in his attempt to construct such policies. By this I do not imply that discounting should not be used in these circumstances, only that I do not know how it should be done. This will be particularly relevant to breeders in a competitive market, since their object will be to sell as many sires as they can. But when sires are sold, their use is controlled by the producer. The breeder would like to sell sires to the producer every year, and would therefore like the buyer to believe that presently available sires are sufficiently better than those bought last year to justify their replacement. Seldom will genetic progress in large animals be rapid enough to justify annual turnover of sires by the producer, but the breeder will be keen to have as quick a turnover as possible. If buyers discount future returns this will be an incentive to the breeder to concentrate on those traits which give quick returns to the producer. Thus in principle discounting of cash flows seems to be the correct procedure.

VI How Detailed Should the Objective Be?

Hill (1981) suggested that the objective be defined simply as profit, and that genetic covariances be calculated between profit and possible selection criteria for the purpose of constructing selection indices. This approach has much in its favour, especially simplicity. It is closely related to the method of comparing strains in poultry random tests by income over feed costs. In some cases it would be very difficult to apply. Grazing animals are difficult to fit into this context since the major cost item, feed, cannot be measured with precision for individuals. Furthermore, when relative prices change, the profit from different animals also changes, so that historical income and expenditure may no longer be relevant. All genetic covariances with the objective would need to be reestimated. We would also, I believe, not have so clear an understanding of why certain criteria were or were not important for selection.

On the other hand, we clearly should not try to specify the objective in too detailed a fashion. For an extreme example, we do not need to specify an economic weight for every gene locus. This is just as well, since we could not do this.

My own opinion is that we should specify each item of income and expenditure. For instance, in sheep this would involve such items as labour for specific purposes (e.g., shearing), veterinary costs, income from sale of weaners, wool and cast for age sheep. It may be tempting to express each of these items as functions of a few traits which affect them. An example of this might be the expression of shearing costs as a function of body size, fleece weight and skin fold. However, this approach has its dangers, since, for instance, skin fold may also affect costs through influencing susceptibility to fly strike. Can we be sure that the level of analysis will be similar for all items and that, in the example, all effects of skin fold are properly accounted for? I think it is difficult to answer this question affirmatively.

At the risk of repeating myself too often, let me say that I believe this process of analysing monetary items into functions of specific traits is often due to a failure to avoid confusion between objectives and criteria.

VII Conclusion

In this paper I have not attempted a complete survey of the field. Pearson and Miller (1981) have recently given a review of what has been done in dairy cattle and, except for details peculiar to the industry, what they have reviewed has general application. Rather, I have tried to select what are unsettled topics, and give some attention to these. I think it appears clearly enough that there is no straight forward answer to the question: How should I define my breeding objective? A variety of methods is possible and, depending on circumstances, one or other of these might be chosen. I have tried to point out what I believe to be problems which have not always been recognised. However, it does seem likely that there will continue to be a degree of divergence in definition of objectives. This may be no more than a reflection of different fundamental aims, rather than being caused by use of faulty methods.

While I believe it is important to make every effort to define our objectives as clearly as possible, a number of studies (for example, that of Ponzoni and Walkley, 1981) have shown that quite large changes in objective may not greatly affect the rate of genetic progress obtained. In view of the difficulties of definition mentioned above, this is a comforting result.

REFERENCES

Andrus, D.F., and McGilliard, L.D. (1975). *J. Dairy Sci. 58,* 1876.

Dickerson, G.E. (1970). *J. Anim. Sci. 30,* 849.

Dickerson, G.E. (1976). *In* "Meat Animals: Growth and Productivity" (D. Lister, D.N. Rhodes, V.R. Fowler and M.F. Fuller, eds.), p. 449. Plenum, New York.

Fowler, V.R., Bichard, M., and Pease, A. (1976). *Anim. Prod. 23,* 365.

Harris, D.L. (1970). *J. Anim. Sci. 30,* 860.

Hill, W.G. (1981). *Proc. Second Conf. Aust. Assoc. Anim. Breed. Genet.* p. 227.

James, J.W. (1978). *Anim. Prod. 26,* 111.

James, J.W. (1980). *Ann. Génét. Sél. anim. 12,* 33.

James, J.W. (1981). *Proc. XIV Int. Congr. Genet.* Vol. 1, Book 2, p. 221.

Land, R.B. (1981). *Livest. Prod. Sci. 8,* 96.
McClintock, A.E., and Cunningham, E.P. (1974). *Anim. Prod. 18,* 237.
Melton, B.E., Heady, E.O., and Willham, R.L. (1979). *Anim. Prod. 28,* 279.
Moav, R. (1973). *In* "Agricultural Genetics" (R. Moav, ed.), p. 319. Wiley, New York.
Morris, C.A. (1981). *Aust. J. exp. Agric. Anim. Husb. 21,* 464.
Pearson, R.E., and Miller, R.H. (1981). *J. Dairy Sci. 64,* 857.
Ponzoni, R.W., and Walkley, J.R.W. (1981). *Livest. Prod. Sci. 8,* 331.
Thompson, R. (1980). *Anim. Prod. 31,* 115.
Young, S.S.Y., and Tallis, G.M. (1961). *J. Anim. Sci. 20,* 506.

Economic Aspects of Developing Breeding Objectives: A Specific Example; Breeding Objectives for Merino Sheep[1]

L. P. Jones

Department of Agriculture
Animal Research Institute
Werribee, Victoria

A method of developing selection indexes is presented. Likely changes in feed requirements that occur when we change particular characters are taken into account.

The method assumes that we wish to maintain stocking pressure at its present level, so numbers are adjusted as feed requirements per head change.

When the method was applied to selection programs to improve Merino sheep, greater emphasis was given to fleece weight and fibre diameter relative to liveweight and reproduction rate than when the effect on feed consumption was ignored.

Selection indexes are given for a range of conditions.

The index computed by this method was compared to an index using a more correct method where feed consumption was treated as part of the breeding objective and its genetic covariances with selection criteria were included in the matrix equation.

I Introduction

James (1982) has discussed in general principles involved in developing breeding objectives. Several authors (e.g., Turner, 1979) have listed the characters of economic importance for Merino sheep. Ponzoni (1979) defined the importance

[1]*Supported by the Wool Research Trust Fund.*

FUTURE DEVELOPMENTS IN THE
GENETIC IMPROVEMENT OF ANIMALS
ISBN 0 12 078830 6

of different characters quantitatively for a number of situations. I will apply a slightly different approach to that of Ponzoni in defining economic weights. In particular, I will take more account of the increased feed requirements that may result from changes in liveweight and reproductive rate.

Two approaches will be used to derive the economic weight of a particular character. The first is a static approach that compares the economic returns from two flocks which differ for the character of interest.

The second is a dynamic approach which takes account of the time scale of the breeding program. With the dynamic approach, characters increasing net return early in the program become relatively more important (James, 1978). In particular, it can take account of changes in net return in the current generation.

The static approach is considered in detail while the dynamic approach is considered briefly to check if it changes the economic weights appreciably.

James (1982) pointed out the need to distinguish between breeding objectives and selection criteria. In particular, he noted how this confusion occurs in considering feed consumption in selection indexes. He emphasized that we should put an economic weight on feed consumption rather than reduce the returns in particular characters to allow for increased feed consumption. In computing a selection index, the genetic covariances of feed consumption with selection criteria are taken into account.

In this paper, I have used the approach of reducing the returns to allow for the extra feed consumed, even though this approach is not strictly valid. I have done this because it is easier to make realistic guesses of the regression of feed consumed on the level of a character, rather than to estimate or guess the genetic covariance between the selection criteria and the total amount of feed consumed by a flock. It should be noted that this covariance is a function of flock structure. Later in the paper, I discuss the effect of errors in my approach on the selection index.

II Static Approach

The basic approach used in the paper is to assume that the producer wishes to maintain stocking pressure constant. The producer may need to change numbers as some characters change. The effect of relaxing this assumption is assessed later.

A. Effect on Number of Stock

The net return (W) per annum from a flock of N sheep can be written as:

$$W = N(w - c) - C_o$$

where w is the gross return per head,
c is the total cost per head,
C_o is the overhead cost for the flock.

If we change the average breeding value (g_i) for a character by a small amount, the change in W is given by:

$$\frac{\partial W}{\partial g_i} = N\left(\frac{\partial w}{\partial g_i} - \frac{\partial c}{\partial g_i}\right) + \frac{\partial N}{\partial g_i}(w - c)$$

plus second order terms which can be neglected without serious loss in accuracy.

In this and later equations, we are not strictly using partial derivatives as, for example, N is not being held constant. All g_j for $j \neq i$ are held constant.

If the amount eaten is proportional to $g_i^{X_i}$ where X_i is a constant,

$$N = k\, g_i^{X_i} \quad \text{where } k \text{ is a constant, and}$$

$$\frac{\partial N}{\partial g_i} = -X_i\, k\, g_i^{-X_i}\, g_i^{-1}$$

$$= -N\, X_i / g_i$$

Thus the change in W will be a balance between the increased net return per head and the decrease in the number of animals that can be carried.

B. Return per Head

Where several products are sold, the net return per head is given by:

$$w - c = \sum_i \left[f_i (y_i p_i - c_i) \right] - c_o$$

where f_i is the fraction of population giving returns for the character i,
y_i is the mean for character i,
p_i is the price of product i,
c_i is the cost associated with character i,
c_o is the general per head cost.

If the mean breeding value (g_i) of a character is changed by a small amount (Δg_i) the change in net return per head is given by:

$$\frac{\partial w}{\partial g_i} - \frac{\partial c}{\partial g_i} = \sum_j \left[f_j \left\{ \frac{\partial y_j}{\partial g_i} p_j + y_j \frac{\partial p_j}{\partial g_i} - \frac{\partial c_j}{\partial g_i} \right\} + \frac{\partial f_j}{\partial g_i} (y_j p_j - c_j) \right]$$

The first term accounts for the extra production from character j from change in breeding value of i. Generally, this term is zero for $j \neq i$, but it allows for the fact that increasing the number of lambs born reduces wool weight per head. The second term allows for changes in price per unit product from a change in wool quality or even liveweight. The third term allows for changes in per head costs that result from change in a character. The fourth term accounts for the change in proportion of the population giving returns in one character that results from changes in another character (e.g., lambs weaned). A simpler approach to changes in number of lambs weaned is to compute the value of the extra lambs reared less rearing costs.

C. Example

Consider a flock of 1,000 ewes with 5 age groups. The following assumptions are made:

The number of lambs weaned per ewe joined is 0.7 for 2-year-old ewes, 0.9 for older ewes.
The death rate after weaning is 5% per annum.
Surplus lambs are sold after shearing at 16 months.
Mean hogget liveweight (HLW) = 40 kg.
Mean adult liveweight (ELW) = 50 kg.
Mean hogget clean fleece weight (HFW) = 4 kg
Mean adult clean fleece weight (EFW) = 4 kg.
Mean lamb clean fleece weight = 1.7 kg.

Prices for the characters are:
Hogget liveweight 40¢/kg irrespective of weight.
Adult liveweight 30¢/kg.
Clean wool weight 300¢/kg.
The margin per micron is 10¢/kg.

The numbers in each age group at mating are:

	Lambs	$1\frac{1}{2}$	$2\frac{1}{2}$	$3\frac{1}{2}$	$4\frac{1}{2}$	$5\frac{1}{2}$	$6\frac{1}{2}$
Breeding	0	221	210	200	189	180	
Non-breeding	856	592					171

Return per ewe per year includes:

Hogget sales	0.592 x 40 x 0.4	= $ 9.50
Cast for age ewes	0.171 x 50 x 0.3	= $ 2.50
Hogget and adult wool sales	1.763 x 4 x 3	= $21.20
Lambs wool	0.856 x 1.7 x 3	= $ 4.30
		$37.50

If we assume that the per head cost is $7.50; the net return per ewe = $30.

The economic weight of a character is the change in net return per unit change in the character. The return will be considered per initial ewe per year. Some (e.g., Ponzoni, 1979) use the return per ewe lifetime, but the time span does not really make any difference. The ewe numbers may have to change as characters change. I will be considering the change in net return divided by 1,000 to account for such changes.

The characters considered in the breeding objective are: Weaning weight (WW), hogget liveweight (HLW), adult liveweight (ELW), hogget fleece weight (HFW), adult fleece weight (EFW), fibre diameter (FD), number of lambs born (NLB), number of lambs weaned (NLW). Feed consumption is also a character in the breeding objective. For the reasons given earlier I have chosen to reduce the returns from the other characters, even though this is not strictly valid.

We now consider the effect of stock numbers and returns per head as the characters change.

1. The Effect of Each Character on Number of Stock $(\frac{\partial N}{\partial g_i})$. I will assume that a ewe and lamb eat 1.7 dry sheep equivalents (D.S.E.) and a hogget 0.9 D.S.E. With the lambing percentage given we would require 0.857 x 1.7 + 0.144 x 1 + 0.813 x 0.9, i.e., 2.330 D.S.E. per ewe. The relative proportions eaten by adults and hoggets is about 0.69:0.31.

a. Weaning weight (WW). As lambs eat a relatively small proportion of the pasture the effect of weaning weight *per se* on the numbers carried will be small. Assume $\partial N/\partial g_1 = 0$.

b. Hogget liveweight (HLW). If the amount eaten is proportional to liveweight$^{0.75}$, a reasonable approximation would be $x_2 = 0.75 \times 0.31$, i.e., $\partial N/\partial g_2/N = -0.23/y_2 = -0.006$.

c. Ewe liveweight (ELW). $x_3 = 0.75 \times 0.69$, $\partial N/\partial g_3/N = -0.010$.

d. Fleece weight and diameter. Changes in fleece weight and diameter will not affect maintenance requirements. The few studies on efficiency of wool production with different strains suggest that most of the differences in wool production are due to differences in efficiency and liveweight, with only a small effect from differences in intake per unit liveweight (Williams, 1979). For a start, I will assume that the effect of wool production and diameter on carrying capacity *per se* is negligible (i.e., $\partial N/\partial g_i = 0$ for HFW, EFW and FD).

e. Number of lambs born (NLB). Each extra lamb born increases the feed requirement by about 50% for six weeks or 6% on an annual basis (MAF, 1975). With adults eating 69%, the increased requirement is 0.041, i.e., $\partial N/\partial g_7/N = -0.041$.

f. Number of lambs weaned (NLW). Each lamb weaned increases the feed requirements of the ewe by about 50% for 4 months or 17% on an annual basis (MAF, 1975), and increases the number of hoggets by 0.95. The number of D.S.E. required is increased by 1.7 x 0.9 x 0.95 = 1.14. The number of stock is reduced by about 0.50 (1.14/2.3), i.e., $\partial N/\partial g_8/N = 0.5$. In other words, for every two extra lambs weaned the number of ewes could be reduced by about one.

2. *Return per Head for Each Character ($\partial w/\partial g_i - \partial c/\partial g_i$).* It is assumed that cost per head is independent of liveweight, fleece weight and diameter, i.e., $\partial c/\partial g_i = 0$ for WW, HLW, HFW, EFW and FD.

WW. Since there are no sales at weaning $\partial w/\partial g_1 = 0$.
HLW. $\partial w/\partial g_2 = 0.592 \times 40 = 23.7$ cents.
ELW. $\partial w/\partial g_3 = 0.171 \times 30 = 5.13$ cents.
HFW. $\partial w/\partial g_4 = 0.813 \times 300 = 244$ cents.
EFW. $\partial w/\partial g_5 = 0.95 \times 300 = 285$ cents.

FD. $\partial w/\partial g_6$ = 1.763 x -40 = -70 cents since each micron finer increases the value per fleece by 40 cents.

NLB. $\partial w/\partial g_7$ = 0 since number of lambs born reduces wool production by about 8%, i.e., $\partial w/\partial g_7 - \partial c/\partial g_7$ = -1200 x 0.95 x 0.08 = -91 cents.

NLW. The extra lambs would mainly be twins with about 5% lower hogget liveweight and fleece weight, and 15% less wool as weaners. With a lamb's fleece worth 500 cents we have:

$$\partial w/\partial g_8 = (1600 + 1200)(2 \times 0.95 - 1) + 500(2 \times 0.85 - 1) = 2870 \text{ cents.}$$

$\partial c/\partial g_8$ The extra lamb would reduce its dam's fleece weight by about 5% and would have extra management costs (drenching, shearing) so total cost might be about 500 cents, i.e., $\partial w/\partial g_8 - \partial c/\partial g_8$ = 2370 cents.

The combined effect of return per head and the change in numbers are shown in Table I. Expressing the economic weights in units of standard deviation gives an indication of their likely importance.

3. Application of the Economic Weights to a Selection Index. The values in Table I suggest that substantial economic gains can be made from changes in NLW, fleece weight (HFW and EFW) and fibre diameter (FD), whereas liveweight has little economic importance. The economic value of a genetic standard deviation ($ah\sigma$) gives a better indication of the potential for economic gains through selection. Values for $ah\sigma$ are also given in Table I. The values of heritability used are given in Appendix 1.

TABLE I. Increased Net Return per Head, the Change in Stock Numbers and Economic Weights of the Characters

Character	$\partial N/\partial g_i/N$	$\frac{\partial w}{g_i} - \frac{\partial c}{g_i}$	*Economic weight (a)*	*Standard dev'n (σ)*	$a\sigma$	$ah\sigma$
WW	0.0	0.0	0	3.0	0	0
HLW	-0.006	23.7	6	4.5	27	15
ELW	-0.010	5.1	-25	4.5	-112	-71
HFW	0.0	244.0	244	0.5	122	77
EFW	0.0	285.0	285	0.5	142	90
FD	0.0	-70.0	-70	2.2	-154	-109
NLB	-0.041	-91.0	-214	0.4	-86	-38
NLW	-0.50	2370.0	870	0.45	392	152

I will consider a case where rams are selected on HLW, HFW, FD and their type of birth (NLB(D)). The index is the solution of the matrix equation:

$$b = P^{-1}\ G\ a$$

where P is the phenotypic covariance among the selection criteria, and G is the genetic covariance of the selection criteria with characters of economic importance. With characters expressed in units of the standard deviation, terms in P are phenotypic correlations, and in G are products of the genetic correlation and the square roots of the heritabilities (i.e., $r_g h_1 h_2$).

The P and G matrices were derived from values of heritabilities, phenotypic and genetic correlations given in Appendix 1. These matrices are given in Tables II and III.

The appropriate index in standardized units is:

$$I_1 = -9\ HLW + 93\ HFW - 75\ FD + 16\ NLB(D)$$

or in actual units:

$$I_1 = 2.1\ HLW(kg) + 185\ HFW(kg) - 34\ FD(\mu m) + 40\ NLB(D)$$

TABLE II. Phenotypic Covariance Matrix (P) Among the Selection Criteria

	HLW	HFW	FD	NLB(D)
HLW	1.0	0.3	0.13	0.028
HFW	0.3	1.0	0.2	0.0
FD	0.13	0.2	1.0	0.0
NLB(D)	0.028	0.0	0.0	1.0

TABLE III. Genetic Covariance Matrix (G) Between Selection Criteria and Characters of Economic Importance

	WW	HLW	ELW	HFW	EFW	FD	NLB	NLW
HLW	0.28	0.40	0.32	0.08	0.08	0.04	0.056	0.061
HFW	0.08	0.08	0.07	0.40	0.32	0.09	0.00	0.00
FD	0.03	0.04	0.04	0.09	0.08	0.50	0.00	0.00
NLB(D)	0.022	0.028	0.028	0.00	0.00	0.00	0.10	0.069

Cunningham (1969) showed that the importance of a character in an index depends on b_i^2/W_{ii} where W_{ii} is the term of P^{-1} appropriate to the character. As the diagonals of P^{-1} are usually close to unity when units of standard deviation are used, b_i^2 gives an indication of the likely importance of the character as a selection criterion.

If we define the relative importance of a character in the index as the percentage reduction in overall economic gain if the character is omitted from the index, the relative importances of the characters are 0.4% for HLW, 42% for HFW, 27% for FD and 1.2% for NLB(D). In other words, HLW and NLB(D) could be omitted with very little reduction in overall gain. The low economic weight for HLW is the reason for its low importance, while selection on NLB(D) is a fairly inefficient way of increasing number of lambs weaned.

4. Varying the Assumptions.

a. Modifying $\partial N/\partial g_i$ for liveweight and reproductive rate. The effect of reproductive rate on stocking rate will be affected by lambing time. For example, if ewes are lactating on lush perennial pastures the extra requirements for improved reproduction may be less than I have assumed. Also, where pastures are understocked, the cost of running larger sheep may be less. However, it would be simpler to increase returns by running more sheep.

Economic weights are given in Table IV for cases where the effect of liveweight and reproductive weight is less than I have suggested. If the values are only half that suggested the index becomes:

$$I_2 = 53\ HLW + 82\ HFW - 77\ FD + 43\ NLB(D)$$

The relative importance of the characters as selection criteria becomes 8% for HLW, 20% for HFW, 19% for FD and 6% for NLB(D).

b. Allowing for the effect of fleece weight on stocking rate. Williams (1979) suggested that genetic increases in wool production per head have arisen largely from increases in efficiency and to a lesser extent from increases in liveweight. Wool growth does not increase maintenance requirements but energy and protein are used in wool production. In the absence of further data we might assume some increase in stocking pressure. If the feed requirements of an animal increase by 0.25% for a 1% increase in wool production, the

TABLE IV. Economic Weights in Units of Standard Deviation if the Effect on Stocking Rate is a Fraction of the Value Assumed in Table I

Character	Fraction of effect assumed in Table I				
	1.0	0.75	0.50	0.25	0.0
HLW	27	47	67	87	107
ELW	-112	-78	-45	-11	23
HFW	122	122	122	122	122
EFW	142	142	142	142	142
FD	-154	-154	-154	-154	-154
NLB	-86	-73	-61	-49	-36
NLW	392	560	729	898	1066

economic weights for HFW and EFW would be reduced to 93 and 78 cents respectively. With a 50% increase, economic weights would be 64 and 13 cents respectively. The relative importance of the characters as selection criteria would be 0.3% for HLW, 23% for HFW, 47% for FD and 1.7% for NLB with a 25% increase and 0.1%, 5%, 73% and 2% for 50% respectively.

c. Relative prices of meat and wool. Clearly increases in the price of meat relative to wool will change the economic weights of the characters. However, even with double the price of meat assumed, fleece weight and fibre diameter are the major selection criteria.

d. Price per kg as a fraction of liveweight. Since the development of the live sheep trade, larger wethers are often worth more per kilogram than small wethers. Even when 50 kg wethers bring twice as much per kg as 40 kg wethers, fleece weight and diameter are the major selection criteria.

e. Varying several assumptions. If we vary several assumptions simultaneously the problem becomes multidimensional. Results for a range of combinations are given in Appendices 2 to 4.

III Dynamic Approach

James (1978) showed how the selection index can be designed when returns from different characters occur at different stages in the life cycle. Returns occurring later are discounted more than those occurring early. It is particularly useful for designing an index to maximize the total benefits from gains in the current and future generations.

For simplicity, only genetic gains will be considered here. The coefficients are the solutions to $b = P^{-1} G A Q c$ (James, 1978) where

A is a matrix describing the return from each group.

Q is a matrix describing the proportion of genes in each age group coming from the selected group for each year of the program.

c is a vector with term c^t which discounts the return in year t to its present value.

James (1978) showed that the problem simplifies if we take the extra returns to infinity from selection in year 0. The solution becomes

$$b = P^{-1} G A \lambda \begin{bmatrix} c \\ c^2 \\ \vdots \\ c^n \end{bmatrix}$$

where λ is a scalar and n is age at which sheep are culled for age. Relative economic weights can be obtained for this situation without computing λ.

The terms in A used by James (1978) were modified slightly since there is an extra lag in the benefits and some costs from reproduction. The benefits from reproduction were discounted by one year to their value at the birth of the lamb. The effects of changes in stock numbers were modified slightly since reductions in stock numbers lead to increased sales in the early stages. The A matrix is given in Table V.

TABLE V. The A *Matrix for the Case Considered*

Character	Age						
	0	1	2	3	4	5	6
HLW	0	30.0					
ELW	0		-31.4	-29.9	-28.5	-26.9	-0.56
HFW	0	122.0					
EFW	0		31.5	30.0	28.4	27.0	25.6
FD	0	-71.5	-18.5	-17.6	-16.6	-15.8	-15.0
NLB	0		-11.2	-10.6	-10.1	-9.6	-9.1
NLW	0		81.9	77.8	74.0	70.0	66.7

With a discount rate of 10%, the $(Qc)'$ vector for selection in males which are used for two years and taking returns to infinity is given by:

$$(Qc)' = 1.668\ (0.909\ 0.826\ 0.751\ 0.683\ 0.621\ 0.564\ 0.513)$$
$$= (1.516\ 1.378\ 1.253\ 1.139\ 1.035\ 0.941\ 0.856)$$
$$(A\,Q\,c)' = (41.3\ -130\ 168\ 181\ -187\ -53\ 391)$$

and the index

$$I_3 = -9.7\ HLW + 123.5\ HFW - 90.8\ FD + 19.4\ NLB(D).$$

This index has a correlation greater than 0.99 with the index I_1 for the static model. In other words making allowance for the effect of time lag had little effect on the relative economic weights in this case. Even if we considered the gains for only 15 years an index with a correlation with I_1 of greater than 0.99 was obtained.

An advantage of the dynamic approach suggested by James (1978) is that it gives a measure of the economic value of the selection program. The standard deviation of the selection index $(b'\,P\,b)^{\frac{1}{2}}$ for gains to time infinity was 138 cents. In other words the present value of genetic gains from selection in rams this year would be 138 cents per unit of standardized selection differential. As gains would be reduced by less than 1% if NLB(D) or HLW were omitted it enables a producer to decide readily if the cost of recording these characters is worthwhile.

It must be realized that the value computed above is only approximate. Hopkins and James (1977) showed that gains could be increased by 10 to 15% if the selection program was modified so that parents were kept for an optimum time and account was taken of the parents' age. Also, in a stud or nucleus scheme the A and Q matrices would need to be adjusted to take account of genetic gains in daughter flocks.

IV Treating Feed Consumption as an Objective

As mentioned by James (1982), the valid approach to considering feed consumption is to include it in the breeding objective. Its genetic covariances with the selection criteria are taken into account in calculating the selection index.

The genetic covariances that would give the identical index to I_1 can readily be calculated. The problem is to find the G matrix and a vector that give identical solutions to $Pb = G\ a$. In other words we solve:

$$G^* a^* = G\ a$$

where G^* and a^* are the values if feed consumption is included in the breeding objective. This can be rewritten as:

$$[G]\ [G_F] \begin{bmatrix} a_O \\ a_F \end{bmatrix} = G\ a$$

or $$[G_F]a_F = [G] \quad [a - a_O]$$

where a_O is the a vector in Table IV for the case with no allowance for feed consumption,

a_F is the economic value of feed consumption.

G_F is the vector of genetic covariances of feed consumption and the selection criteria.

From the values in Table IV, we have:

$$[G_F\ a_F]' = (-119\ -16\ -9\ -58).$$

Assuming a net return of 30 dollars per ewe, a 1% increase in feed consumption costs 30 cents per ewe. With a standard deviation of feed consumption of 15%, $a_F = -450$.

Thus $G_F = (0.26\ 0.04\ 0.02\ 0.13)$.

If the heritability of feed consumption is 0.4, its genetic correlation with the selection criteria would be:

HLW 0.66, HFW 0.09, FD 0.04, and NLB 0.90.

V Discussion

The main conclusion from the study is that we need to take account of the increased feed requirements that changes in characters may cause. The correct approach is to include feed consumption as an objective. However, to develop an index to accommodate increased feed consumption requires knowledge of its genetic covariances with selection criteria. These covariances are a function of flock structure. Estimates are not available for grazing sheep so only guesses can be made. On the other hand, some estimates of the likely change in feed consumption with changes in other characters are available. The approach used here provides a method which gives a selection index which seems reasonable.

The vector of economic weight a can easily be modified to allow for varying assumptions. The values of genetic correlation obtained in Section IV seem reasonable but are only speculative. The main disadvantage of the approach I have used is that it assumes that increased feed consumption is inevitable if we change particular characters. It is at least possible that we can use selection criteria that will increase liveweight without increasing feed consumption. If such criteria could be found my approach would lead to incorrect indexes.

Clearly, I have not produced the ideal index for all situations. The effect of increased reproduction on feed requirements will be less for a producer selling surplus lambs at weaning, especially for a producer who has a lambing season with surplus feed.

Another important consideration is the robustness of the index to assumptions. The correlation r_{12} between indexes provides a measure of robustness. This is given by:

$$r_{12} = b_1{}'P\,b_2 \;/\; \left[(b_1{}'P\,b_1)^{\frac{1}{2}}(b_2{}'P\,b_2)^{\frac{1}{2}}\right]$$

The correlations of I_1 with indexes, appropriate when the effects of stocking pressure on liveweight and reproductive rate were 75% and 50%, were 0.96 and 0.86 respectively, while if each increase of 1% in wool weight leads to declines of 0.25% or 0.5% in stock numbers, I_1 had correlations of 0.96 or 0.80 with the appropriate index.

Actually I_1 had a correlation of 0.98 with an index giving equal weight to fleece weight and diameter. In other words, if I_1 was the appropriate index, selection with equal weight to fleece weight and diameter would be 98% as efficient.

The results suggest that it is not worth selecting for reproduction rate. This conclusion needs to be kept in perspective. The economic weight suggests that increased reproduction rate is worthwhile. However, the criterion used in this example to assess reproduction rate is not an accurate assessment of breeding value for number of lambs weaned. Where information on several lambings is available the weight given to reproduction would increase. An alternative approach would be to make use of between flock differences where these are known. The value will also depend on the relative survival of twins and singles in an environment as well as penalties against twins for fleece weight.

Again, it is unlikely that selection for increased liveweight is worthwhile in most situations. However, its benefit will be increased where increased liveweight provides new potential sales. This occurs to some extent in the live sheep trade where larger animals are preferred. Also, with larger types it may be possible to sell more lambs at or soon after weaning. It would be ideal if we could develop sheep with greater appetites in periods of plenty but which have no greater requirements when feed was scarce.

In this study, I have illustrated how economic weights can be determined. The approach can be applied to particular situations and can be extended to cover other characters such as resistance to fly-strike.

REFERENCES

Cunningham, E.P. (1969). *Acta Agric. scand. 19*, 45.

Hopkins, I.R., and James, J.W. (1977). *Anim. Prod. 25*, 111.

James, J.W. (1978). *Anim. Prod. 26*, 111.

James, J.W. (1982). Chapter 6, this Volume.

MAF (1975). Ministry of Agriculture, Fisheries and Food, Technical Bulletin 33.

Ponzoni, R.W. (1979). *Proc. Aust. Assoc. Anim. Breed. Genet. 1*, 320.

Turner, H.N. (1979). *In* "Sheep Breeding" (G.J. Tomes, D.E. Robertson, R.J. Lightfoot, W. Haresign, eds.), p. 93. Butterworths, London.

Williams, A.J. (1979). *In* "Physiological and Environmental Limitations to Wool Growth" (J.L. Black and P.J. Reis, eds.), p. 337. University of New England, Armidale.

APPENDIX 1. Phenotypic and Genetic Parameters for the Characters Used in the Example

	WW	HLW	ELW	HFW	EFW	FD	NLB	NLW
Standard deviation	3.0	4.5	4.5	0.5	0.5	2.2	0.4	0.45
Heritability	0.25	0.4	0.4	0.4	0.4	0.5	0.2	0.15
				Correlations[a]				
WW		0.35	0.35	0.2	0.2	0.1	0.12	0.12
HLW	0.9		0.7	0.3	0.2	0.13	0.12	0.15
ELW	0.8	0.8		0.2	0.3	0.13	0.12	0.15
HFW	0.25	0.2	0.18		0.6	0.2	0.0	0.0
EFW	0.20	0.2	0.2	0.8		0.2	0.0	0.0
FD	0.08	0.09	0.09	0.2	0.18		0.0	0.0
NLB	0.2	0.2	0.2	0.0	0.0	0.0		0.7
NLW	0.2	0.25	0.25	0.0	0.0	0.0	0.8	

[a]Phenotypic correlations above the diagonal, genetic correlations below the diagonal.

APPENDIX 2. Some Conditions for Which Selection Indexes Were Computed

Condition	Description
1	As in original example.
2	The effect of liveweight (LW) and reproduction rate (RR) on feed requirements half that for 1.
3	No effect of LW or RR on feed required.
4	As for 1, except for 1% increase in fleece weight (FW) increases feed requirements by 25%.
5	Effect of LW and RR as in 2, FW as in 4.
6	As in 1, with price per kg of LW doubled.
7	As in 6, with effect of LW and RR as in 2.
8	As in 6, with effect of FW as in 4.
9	As in 1, with price per kg of LW increased by 2¢/kg for each increase to 1 kg in LW.
10	As in 9, with effect of LW and RR as in 2.
11	As in 10, with effect of FW as in 4.
12	As in 1, with dynamic approach.

APPENDIX 3. *Economic Weights in Standard Deviations Corresponding to Conditions Described in Appendix 2*

	Character						
Condition	HLW	ELW	HFW	EFW	FD	NLB	LBW
1	27	-112	122	142	-154	-86	392
2	67	-45	122	142	-154	-61	729
3	107	23	122	142	-154	-36	1066
4	27	-112	93	78	-154	-86	392
5	67	-45	93	78	-154	-61	729
6	100	-143	112	142	-154	-105	770
7	156	-48	122	142	-154	-71	1242
8	100	-143	81	52	-154	-105	770
9	244	-112	122	142	-154	-86	225
10	284	-45	122	142	-154	-61	592
11	244	-112	93	78	-154	-86	255
12	41	-130	168	181	-187	-53	391

APPENDIX 4. *Coefficients in Standard Deviations of Selection Index Appropriate for each Condition Considered in Appendix 2 (Relative Importance[a] in Brackets)*

	Character				
Condition	HLW	HFW	FD	NLB(D)	Correlation with I_1
1	-9(0.4)	93(42)	-75(27)	16(1.2)	1.00
2	53(8)	82(20)	-77(19)	43(6)	0.86
3	114(22)	72(8)	-79(10)	70(9)	0.65
4	-7(0.3)	60(23)	-77(47)	16(2)	0.96
5	55(11)	50(9)	-79(27)	43(8)	0.79
6	34(4)	84(25)	-78(23)	40(6)	0.91
7	121(23)	69(7)	-80(10)	78(10)	0.63
8	38(7)	38(7)	-80(38)	40(9)	0.80
9	69(13)	86(21)	-76(17)	9(0.2)	0.81
10	133(29)	75(8)	-78(9)	38(2)	0.62
11	74(20)	53(10)	-77(24)	11(0.4)	0.74
12	-10(0.2)	124(46)	-91(23)	19(1.0)	1.00

[a] *The reduction in the response for overall economic merit if the character is omitted from the index*

Discussion Summary

The calculation of economic weights for traits included in a Breeding Objective was seen as being of fundamental importance, although often poorly understood.

The main tenets of Selection Index Theory and the calculation of economic weights have been well documented in the paper by James, but practical problems faced when calculating selection indices were identified in the discussions, and are outlined below.

When an individual farmer is selling a product (e.g., milk) where market interference exists, it is possible for him to calculate the net economic value of increased production for his enterprise in isolation, but his calculations obviously will depend on the exact market constraints operating. When equivalent net economic values are estimated nationally, it is better to use the Export Value of the product.

There are several facets of the problem of non-linearity which should be considered when dealing with selection indices. The first relates to non-linear genetic and phenotypic relationships between the traits themselves. The second relates to non-linear relationships between time and the economic returns per unit product. The latter is at least partially catered for by discounting future returns, thus placing less emphasis on future returns whose realization may not be so assured. The former problem is generally avoided by assuming linearity for small ranges of values. A similar argument is normally applied to traits where an intermediate level has a higher economic value than either of the extremes.

Small errors in economic weights will not have great effects as long as the economically most important traits are given heaviest weighting and their sign (positive or negative) is correct. In cases where economic weights of two traits in the Breeding Objective have opposite signs and the genetic correlation between these traits is positive, the precision of economic weights becomes more important. Similarly, when economic weights have the same sign but the traits have a negative genetic correlation, precision of the weights and the correlations become more important.

Economic weights for sex-limited traits in the Breeding Objective are independent of the sex of the animal being selected, providing returns are recouped *only* in the descendants of the selected individuals. When returns are expressed both by the descendants and by the selected animals, economic weights will usually differ depending on the sex of the individuals being selected.

The economic weights for traits such as temperament or conformation sometimes can be estimated by analysing culling data using a multiple regression model which includes all traits of interest. In the case of dairy cows, one could assess the importance of poor temperament relative to poor production in determining the probability of survival.

Part 6
Mixed Model Theory

The Use of Mixed Model Theory in the Estimation of Parameters

Robert D. Anderson

Department of Sheep Husbandry
Massey University
Palmerston North, New Zealand

I Introduction

Animal breeding data analyses frequently involve mixed linear models, that is, linear models including fixed and random effects. Whereas estimation of parameters of mixed models with balanced data (equal numbers of observations in the sub-classes) is straightforward, estimating parameters, particularly variance parameters, in mixed models with unbalanced data (unequal numbers of observations in the sub-classes, including missing sub-classes) is usually a complicated undertaking. The text of this paper focusses attention on the latter problem.

The paper by C.R. Henderson (1953) represents the first major contribution to the estimation of variance parameters in mixed models with unbalanced data. Henderson's Methods II and III in that paper were designed specifically to furnish unbiased estimates of variance components in mixed models with unbalanced data. In fact, despite the availability of alternative techniques, the characteristically large numbers of observations that occur in animal breeding studies have led to an almost exclusive use of Henderson's methods. Equations which, when solved iteratively, generate maximum likelihood (ML) estimates of the parameters in a 1-way random effects model with unbalanced data, were derived by Crump (1947), but the current interest in using ML for mixed models with unbalanced data is due to Hartley and J.N.K. Rao (1967). They derive ML equations in full

FUTURE DEVELOPMENTS IN THE
GENETIC IMPROVEMENT OF ANIMALS
ISBN 0 12 078830 6

generality, and in doing so, maximize the likelihood over the entire sample space of the observation vector. Extending the study of W.A. Thompson (1962), the particularly important paper of Patterson and R. Thompson (1971) sets out an alternative ML procedure for estimating variance parameters, subsequently known as restricted maximum likelihood (REML), whereby the additional need to estimate fixed effects in a mixed model is taken into account. (It is restricted in the sense that the likelihood is maximized over the restricted population of samples defined by the realized vector of fixed effects; see Kendall 1946; pp 34-46.)

Commencing with the study of Townsend (1968) (also Townsend and Searle, 1971) much attention has been given to the development of quadratic estimators of variance parameters that are unbiased and have minimum sampling variance properties. This led to the technique of minimum norm quadratic unbiased estimation (MINQUE), introduced by C.R. Rao (1970). Assuming normality, La Motte (1970) developed locally minimum variance quadratic unbiased estimators (MIVQUE) of variance parameters. Further MINQUE and MIVQUE theory was elaborated by C.R. Rao (1971a, 1971b, 1972). The dependence of MINQUE and MIVQUE on an *a priori* knowledge of the variance parameters to be estimated may be alleviated by using iterative- or I-MINQUE, but the properties of MINQUE and MIVQUE, as such, are lost in the process.

Now that there is a variety of methods for estimating variance parameters of mixed models with unbalanced data, choosing a method in a given application is a problem facing users in disciplines such as animal breeding, particularly in view of future computing capabilities. The detailing of relationships amongst the methods and the provision of guidance on the appropriateness of the methods in different settings has received insufficient attention. Moreover, the application of some methods has been hindered by the use in their derivation of mathematical techniques that are unfamiliar to researchers in applied disciplines; statistical procedures are vulnerable to mis-use if users do not have some knowledge of the derivations of the procedures.

This paper outlines a unified approach to the derivation of several methods of variance parameter estimation, especially those of recent origin. The concept of dispersion-mean correspondence in linear models (Pukelsheim, 1976) is used to relate variance parameter estimation to well-known, and easily understood, linear estimation theory. The relationships among methods are exhibited clearly by this approach.

II Notation and Mathematical Preliminaries

The mixed linear model assumed in this paper is

$$y = X\beta + \sum_{i=1}^{k} Z_i u_i , \tag{1}$$

where y is an N x 1 vector of observations, X of order N x p and Z_i of order N x c_i, i=1, ..., k, are known matrices, β is an unknown vector of p fixed effects, and u_i, of order c_i x 1, i=1, ..., k, are non-observable vectors of random effects, such that: (i) $E(u_i) = 0$, i=1, ..., k, (ii) the elements of u_i are independently distributed with common variance σ_i^2 and kurtosis γ_i, where $\gamma_i = E(u_{ij}^4)/\sigma_i^4 - 3$, for j=1, ..., c_i, and (iii) u_i and $u_{i'}$ are independent for i≠i'.
The vector of observations, y, has mean and variance

$$E(y) = X\beta \quad \text{and} \quad \text{var}(y) = \sum_{i=1}^{k} \sigma_i^2 V_i = V , \tag{2}$$

where $V_i = Z_i Z_i'$.
An alternative representation of (1) is

$$y = X\beta + Zu, \text{ with } V = ZDZ' \text{ and } D = \sum_{i=1}^{k}{}^{+} \sigma_i^2 I_{c_i} , \tag{3}$$

where $Z = [Z_1 | \ldots | Z_k]$, $u' = [u_1' | \ldots | u_k']$ and Σ^+ denotes a direct sum.

In order to understand variance component estimation, some acquaintance with quadratic forms and functions thereof is needed. In particular, the matrix operator vec which affords considerable convenience in the algebra of quadratic forms will be used extensively. The vec of an r x c matrix B, with columns b_i, i=1, ..., c, is defined to be

$$\text{vec}B = \begin{bmatrix} b_1 \\ b_2 \\ \vdots \\ b_c \end{bmatrix} , \tag{4}$$

that is, the rc-column vector obtained by placing the columns of B one beneath the other (the $(i+1)^{th}$ beneath the ith). Properties of the vec operator are fully detailed in the paper by H.V. Henderson and Searle (1979), those of importance for this paper being: (i) For matrices A, B and C conformable for multiplication in the stated order,

$$\text{vec}(\underset{\sim}{A}\underset{\sim}{B}\underset{\sim}{C}) = (\underset{\sim}{C}' * \underset{\sim}{A})\text{vec}\underset{\sim}{B} \ , \qquad (5)$$

where '*' denotes (strictly) the right Kronecker product of the matrices concerned; (ii) The trace of a matrix product $\underset{\sim}{A}\underset{\sim}{B}$ is (see also Neudecker, 1969)

$$\text{tr}(\underset{\sim}{A}\underset{\sim}{B}) = (\text{vec}\underset{\sim}{A}')'\text{vec}\underset{\sim}{B} \ , \qquad (6)$$

and; (iii) The vec of a matrix $\underset{\sim}{A}_{rxc}$ and the vec of its transpose $\underset{\sim}{A}'_{cxr}$ are related through the identity

$$\text{vec}\underset{\sim}{A}_{rxc} = \underset{\sim}{I}_{(r,c)}\text{vec}(\underset{\sim}{A}') \ , \qquad (7)$$

where $\underset{\sim}{I}_{(r,c)}$ is a 'vec-permutation' matrix so named by H.V. Henderson and Searle (1979).

Consider now the quadratic form $(\underset{\sim}{y} - \underset{\sim}{X}\underset{\sim}{\beta})'\underset{\sim}{A}(\underset{\sim}{y} - \underset{\sim}{X}\underset{\sim}{\beta})$ for some symmetric matrix $\underset{\sim}{A}$. Through being a scalar, it follows that

$$(\underset{\sim}{y} - \underset{\sim}{X}\underset{\sim}{\beta})'\underset{\sim}{A}(\underset{\sim}{y} - \underset{\sim}{X}\underset{\sim}{\beta}) = \{\text{vec}[(\underset{\sim}{y} - \underset{\sim}{X}\underset{\sim}{\beta})'\underset{\sim}{A}(\underset{\sim}{y} - \underset{\sim}{X}\underset{\sim}{\beta})]\}' \ .$$

Applying (5) means that

$$\begin{aligned}(\underset{\sim}{y} - \underset{\sim}{X}\underset{\sim}{\beta})'\underset{\sim}{A}(\underset{\sim}{y} - \underset{\sim}{X}\underset{\sim}{\beta}) &= [(\underset{\sim}{y} - \underset{\sim}{X}\underset{\sim}{\beta})'*(\underset{\sim}{y} - \underset{\sim}{X}\underset{\sim}{\beta})'\text{vec}\underset{\sim}{A}]' \\ &= (\text{vec}\underset{\sim}{A})'[(\underset{\sim}{y} - \underset{\sim}{X}\underset{\sim}{\beta})*(\underset{\sim}{y} - \underset{\sim}{X}\underset{\sim}{\beta})] \ . \qquad (8)\end{aligned}$$

That is, the vec operator is a convenient device for writing quadratic forms in the elements of the vector $\underset{\sim}{y} - \underset{\sim}{X}\underset{\sim}{\beta}$ as linear forms in the elements of the vector $(\underset{\sim}{y} - \underset{\sim}{X}\underset{\sim}{\beta})*(\underset{\sim}{y} - \underset{\sim}{X}\underset{\sim}{\beta})$. Since the principles of linear estimation theory are well understood by animal breeders, this result will be exploited to advantage in what follows. For instance, the transformation leads straightforwardly, as shown by R.D. Anderson, Quaas and Searle (1977) and Pukelsheim (1977), to a general expression for the variance of the abovementioned quadratic form, namely;

$$\begin{aligned}\text{var}[(\underset{\sim}{y} - \underset{\sim}{X}\underset{\sim}{\beta})'\underset{\sim}{A}(\underset{\sim}{y} - \underset{\sim}{X}\underset{\sim}{\beta})] &= (\text{vec}\underset{\sim}{A})'\text{var}[(\underset{\sim}{y} - \underset{\sim}{X}\underset{\sim}{\beta})*(\underset{\sim}{y} - \underset{\sim}{X}\underset{\sim}{\beta})]\text{vec}\underset{\sim}{A} \\ &= (\text{vec}\underset{\sim}{A})'\text{var}[\text{vec}(\underset{\sim}{y} - \underset{\sim}{X}\underset{\sim}{\beta})(\underset{\sim}{y} - \underset{\sim}{X}\underset{\sim}{\beta})']\text{vec}\underset{\sim}{A} \\ &= (\text{vec}\underset{\sim}{A})'\underset{\sim}{F}\ (\text{vec}\underset{\sim}{A}) \ , \qquad (9)\end{aligned}$$

where

$$\underset{\sim}{F} = (\underset{\sim}{V}*\underset{\sim}{V})(\underset{\sim}{I}_{N^2} + \underset{\sim}{I}_{(N,N)}) + (\underset{\sim}{Z}*\underset{\sim}{Z})\text{dg}\{\text{vec}(\sum_{i=1}^{k+}\gamma_i\sigma_i^4\underset{\sim}{I}_{c_i})\}(\underset{\sim}{Z}*\underset{\sim}{Z})' \qquad (10)$$

and where dg{b} signifies a diagonal matrix with elements of $\mathbf{b}$ along the diagonal. In the case of normality, the kurtosis parameters are zero and (10) reduces to

$$\mathbf{F} = (\mathbf{V}*\mathbf{V})(\mathbf{I}_{N^2} + \mathbf{I}_{(N,N)}) \ . \tag{11}$$

III Minimum Variance Quadratic Unbiased Estimators

In this section quadratic estimators of variance parameters, pertaining to the mixed model (1), which are unbiased and which have minimum variance properties will be derived using linear estimation theory.

Denoting the vector of variance parameters by $\boldsymbol{\sigma}^2$, the estimators of the estimable function $\mathbf{t}'\boldsymbol{\sigma}^2$ to be derived are of the form $\mathbf{y}'\mathbf{A}\mathbf{y}$, where the symmetric matrix $\mathbf{A}$ is determined according to three criteria.

(i) Unbiasedness. The expected value of $\mathbf{y}'\mathbf{A}\mathbf{y}$ is

$$E(\mathbf{y}'\mathbf{A}\mathbf{y}) = tr(\mathbf{A}\mathbf{V}) + (\mathbf{X}\boldsymbol{\beta})'\mathbf{A}(\mathbf{X}\boldsymbol{\beta}) = \sum_{i=1}^{k} \sigma_i^2 \, tr(\mathbf{A}\mathbf{V}_i) + \boldsymbol{\beta}'\mathbf{X}'\mathbf{A}\mathbf{X}\boldsymbol{\beta} \ . \tag{12}$$

Thus $E(\mathbf{y}'\mathbf{A}\mathbf{y}) = \mathbf{t}'\boldsymbol{\sigma}^2$ for all $\boldsymbol{\beta}$ and $\boldsymbol{\sigma}^2$ if and only if

$$\mathbf{X}'\mathbf{A}\mathbf{X} = \mathbf{0} \quad \text{and} \quad tr(\mathbf{A}\mathbf{V}_i) = t_i, \ i=1, \ldots, k \ . \tag{13}$$

(ii) Translation invariance. The quadratic form $\mathbf{y}'\mathbf{A}\mathbf{y}$ is said to be translation invariant if, for all vectors $\boldsymbol{\beta}$,

$$\mathbf{y}'\mathbf{A}\mathbf{y} = (\mathbf{y} - \mathbf{X}\boldsymbol{\beta})'\mathbf{A}(\mathbf{y} - \mathbf{X}\boldsymbol{\beta}), \text{ implying (for } \mathbf{A}=\mathbf{A}'\text{) } \mathbf{A}\mathbf{X} = \mathbf{0} \tag{14}$$

(iii) Minimum Variance. In addition to satisfying (13) and (14), the matrix $\mathbf{A}$ is to be determined such that the variance of the estimator of $\mathbf{t}'\boldsymbol{\sigma}^2$ is minimized.

Turning to the need for translation invariance, any quadratic form in $\mathbf{y}$ is translation invariant if and only if it can be expressed as a quadratic form in $\mathbf{Q}\mathbf{y}$ where $\mathbf{Q}$ is any matrix such that $\mathbf{Q}\mathbf{X} = \mathbf{0}$ and $r(\mathbf{Q}) = N - r(\mathbf{X}) = g$ (Kleffe, 1976). Amongst matrices $\mathbf{Q}$, one choice is the matrix $\mathbf{K}$ such that

$$\mathbf{K}\mathbf{X} = \mathbf{0}, \ \mathbf{K}\mathbf{K}' = \mathbf{I}, \ \mathbf{K}'\mathbf{K} = \mathbf{M} = \mathbf{I} - \mathbf{X}(\mathbf{X}'\mathbf{X})^{-}\mathbf{X}' \ , \tag{15}$$

where $\mathbf{K}_{g \times N}$ has full row rank g and $(\mathbf{X}'\mathbf{X})^{-}$ is a generalized inverse of $(\mathbf{X}'\mathbf{X})$. Then, the vector $\mathbf{K}\mathbf{y}$ can be viewed as a set of g linearly independent error contrasts in the vector $\mathbf{y}$.

(An error contrast is a linear combination $k'y$ such that $E(k'y) = 0$, that is, $k'X = 0$ and k does not depend on either σ^2 or β in the model.) Thus, a translation invariant quadratic form $y'Ay$ can be written as $y'Ay = y'K'BKy$ for some symmetric matrix B.

Applying the results of the previous section, it follows that the translation invariant quadratic form $y'Ay$ can be represented as

$$y'Ay = (\text{vec}B)'[(Ky)*(Ky)] \,, \tag{16}$$

with expected value being

$$\begin{aligned} E(y'Ay) &= (\text{vec}B)'E[(Ky)*(Ky)] \\ &= (\text{vec}B)'\text{vec}[(E(Kyy'K')] \\ &= (\text{vec}B)'\text{vec}[K\ E((y - X\beta)(y - X\beta)')K'] \\ &= (\text{vec}B)'\text{vec}(KVK) \\ &= (\text{vec}B)'(K*K)\text{vec}V \,, \end{aligned}$$

and on defining $\Psi = [\text{vec}V_1 \,|\ldots|\, \text{vec}V_k]$,

$$E(y'Ay) = E(\text{vec}B)'[(Ky)*(Ky)] = (\text{vec}B)'(K*K)\Psi\sigma^2 \,. \tag{17}$$

The variance of $y'Ay$ in the form of the variance of $(\text{vec}B)'[(Ky)*(Ky)]$ is

$$\begin{aligned} \text{var}(y'Ay) &= \text{var}\{(\text{vec}B)'[(Ky)*Ky)]\} \\ &= (\text{vec}B)'(K*K)\text{var}[(y - X\beta)*(y - X\beta)](K*K)'(\text{vec}B) \\ &= (\text{vec}B)'(K*K)F(K*K)'\text{vec}B \,, \end{aligned} \tag{18}$$

on using (9) and (10).

The point to observe from equations (17) and (18) is that the vector $(Ky)*(Ky)$ has mean vector

$$(K*K)\Psi\sigma^2 = \mathrm{X}\sigma^2, \text{ for } \mathrm{X} = (K*K)\Psi \,, \tag{19}$$

and variance-covariance matrix

$$\Omega = (K*K)F(K*K)'. \tag{20}$$

That is, a derived linear model has been constructed such that the variance (dispersion) parameters from the mixed model (1) form the mean parameters in the derived model. Pukelsheim (1976) has termed this to be dispersion-mean correspondence in linear models. Thus, the task of determining the matrix A in accordance with the criteria (i), (ii) and (iii) is now a matter of determining the linear function such that

$$\text{var}[(\text{vec}\underset{\sim}{B})'(\underset{\sim}{K}\underset{\sim}{y})*(\underset{\sim}{K}\underset{\sim}{y})] \text{ is a min. subject to } \underset{\sim}{t}' = (\text{vec}\underset{\sim}{B})'\underset{\sim}{X}^{\sim}. \tag{21}$$

Before applying the usual G.L.S. linear estimation theory to the derived model

$$(\underset{\sim}{K}\underset{\sim}{y})*(\underset{\sim}{K}\underset{\sim}{y}) \sim (\underset{\sim}{X}\underset{\sim}{\sigma}^2, \underset{\sim}{\Omega}) , \tag{22}$$

a critical point to observe is that the variance-covariance matrix Ω is, through the matrix $\underset{\sim}{F}$, a function of not only the variance parameters to be estimated, $\underset{\sim}{\sigma}^2$, but also, the unknown kurtosis parameters. To overcome this difficulty, assume for the moment that the kurtosis parameters are zero (the case under normality) and that the variance-covariance matrix $\underset{\sim}{V}$ in $\underset{\sim}{\Omega}$ has the pre-assigned value $\underset{\sim}{V}_o$ (positive definite) with $\underset{\sim}{\Omega}_o$ being $\underset{\sim}{\Omega}$ with $\underset{\sim}{V}_o$ replacing $\underset{\sim}{V}$. That is, with $\underset{\sim}{V}_o$ "known" *a priori* and all the kurtosis parameters assumed to be zero, $\underset{\sim}{\Omega}_o$ is the "known" matrix

$$\underset{\sim}{\Omega}_o = (\underset{\sim}{K}*\underset{\sim}{K})[(\underset{\sim}{V}_o*\underset{\sim}{V}_o)(\underset{\sim}{I} + \underset{\sim}{I}_{(N,N)})](\underset{\sim}{K}*\underset{\sim}{K})'$$

which, on noting that $\underset{\sim}{I}_{(N,N)}(\underset{\sim}{K}*\underset{\sim}{K})' = (\underset{\sim}{K}*\underset{\sim}{K})'\underset{\sim}{I}_{(g,g)}$, is

$$\underset{\sim}{\Omega}_o = [(\underset{\sim}{K}\underset{\sim}{V}_o\underset{\sim}{K}')*(\underset{\sim}{K}\underset{\sim}{V}_o\underset{\sim}{K}')](\underset{\sim}{I} + \underset{\sim}{I}_{(g,g)}). \tag{23}$$

With $\underset{\sim}{\Omega}_o^-$ being any generalized inverse of $\underset{\sim}{\Omega}_o$, it is readily shown that $\underset{\sim}{\Omega}_o(\underset{\sim}{\Omega}_o)^-\underset{\sim}{X} = \underset{\sim}{X}$ and, therefore, by Corollaries 1.1 and 1.3 of Zyskind and Martin (1969), it can be asserted that the b.l.u.e. of $\underset{\sim}{t}'\underset{\sim}{\sigma}^2$, under the derived model with $\underset{\sim}{\Omega}_o$ in place of $\underset{\sim}{\Omega}$, is $\underset{\sim}{t}'\dot{\underset{\sim}{\sigma}}^2$ where $\dot{\underset{\sim}{\sigma}}^2$ is a solution to the normal equations

$$(\underset{\sim}{X}'\underset{\sim}{\Omega}_o^-\underset{\sim}{X})\dot{\underset{\sim}{\sigma}}^2 = \underset{\sim}{X}'\underset{\sim}{\Omega}_o^-[(\underset{\sim}{K}\underset{\sim}{y})*(\underset{\sim}{K}\underset{\sim}{y})] . \tag{24}$$

Noting that $(\underset{\sim}{I} + \underset{\sim}{I}_{(g,g)})^2 = 2(\underset{\sim}{I} + \underset{\sim}{I}_{(g,g)})$ one value of $\underset{\sim}{\Omega}_o^-$ is $\frac{1}{4}(\underset{\sim}{I} + \underset{\sim}{I}_{(g,g)})[(\underset{\sim}{K}\underset{\sim}{V}_o\underset{\sim}{K}')^{-1}*(\underset{\sim}{K}\underset{\sim}{V}_o\underset{\sim}{K}')^{-1}]$ (recall that $\underset{\sim}{K}$ has full row rank and $\underset{\sim}{V}_o$ is positive definite) which, with $\underset{\sim}{X}' = \underset{\sim}{X}'\underset{\sim}{I}_{(g,g)}$ means that the normal equations (24) are

$$\underset{\sim}{X}'[(\underset{\sim}{K}\underset{\sim}{V}_o\underset{\sim}{K}')^{-1}*(\underset{\sim}{K}\underset{\sim}{V}_o\underset{\sim}{K}')^{-1}]\underset{\sim}{X}\dot{\underset{\sim}{\sigma}}^2 = \underset{\sim}{X}'[(\underset{\sim}{K}\underset{\sim}{V}_o\underset{\sim}{K}')^{-1}*(\underset{\sim}{K}\underset{\sim}{V}_o\underset{\sim}{K}')^{-1}][(\underset{\sim}{K}\underset{\sim}{y})*(\underset{\sim}{K}\underset{\sim}{y})]. \tag{25}$$

Substituting for $\underset{\sim}{X}$ and observing that

$$\underset{\sim}{K}'(\underset{\sim}{K}\underset{\sim}{V}_o\underset{\sim}{K}')^{-1}\underset{\sim}{K} = \underset{\sim}{V}_o^{-1} - \underset{\sim}{V}_o^{-1}\underset{\sim}{X}(\underset{\sim}{X}'\underset{\sim}{V}_o^{-1}\underset{\sim}{X})^{-}\underset{\sim}{X}'\underset{\sim}{V}_o^{-1} = \underset{\sim}{P}_o \,, \quad (26)$$

equations (25) are

$$\{(\text{vec}\underset{\sim}{V}_i)'(\underset{\sim}{P}_o*\underset{\sim}{P}_o)\text{vec}\underset{\sim}{V}_j\}\dot{\underset{\sim}{\sigma}}^2 = \{(\text{vec}\underset{\sim}{V}_i)'(\underset{\sim}{P}_o\underset{\sim}{y}*\underset{\sim}{P}_o\underset{\sim}{y})\} \text{ for } i,\ j=1,\ \ldots,\ k. \quad (27)$$

Applying equation (6), the normal equations under the derived model with $\underset{\sim}{\Omega} = \underset{\sim}{\Omega}_o$ are

$$\{tr(\underset{\sim}{P}_o\underset{\sim}{V}_i\underset{\sim}{P}_o\underset{\sim}{V}_j)\}\dot{\underset{\sim}{\sigma}}^2 = \{\underset{\sim}{y}'\underset{\sim}{P}_o\underset{\sim}{V}_i\underset{\sim}{P}_o\underset{\sim}{y}\},\ i,\ j=1,\ \ldots,\ k, \quad (28)$$

which are identically the MINQUE equations (C.R. Rao 1971a, C.R. Rao 1972). Thus, as Pukelsheim (1976) indicated, applying generalized least squares under the derived model with $\underset{\sim}{\Omega} = \underset{\sim}{\Omega}_o$, gives MINQUE of $\underset{\sim}{\sigma}^2$ under the mixed model (1). In short, the MINQUE of $\underset{\sim}{\sigma}^2$ (note the assumptions regarding $\underset{\sim}{\Omega}$) is a solution to the minimization problem defined by the aforementioned criteria (i), (ii) and (iii).

At this stage it is pertinent to focus attention on the properties of $\underset{\sim}{t}'\dot{\underset{\sim}{\sigma}}^2$ where $\dot{\underset{\sim}{\sigma}}^2$ is a solution to equations (28). It has been established that $\underset{\sim}{t}'\dot{\underset{\sim}{\sigma}}^2$ is the b.l.u.e. of $\underset{\sim}{t}'\underset{\sim}{\sigma}^2$ under the derived model with $\underset{\sim}{\Omega}$ assumed to be equal to $\underset{\sim}{\Omega}_o$, but what can be inferred about $\underset{\sim}{t}'\dot{\underset{\sim}{\sigma}}^2$ in the absence of any assumptions concerning the variance-covariance matrix $\underset{\sim}{\Omega}$ (more specifically assumptions concerning $\underset{\sim}{V}$ and the kurtosis parameters)? In the special case when

$$\underset{\sim}{\Omega}\underset{\sim}{X} = \underset{\sim}{X}\underset{\sim}{C} \,, \quad (29)$$

for some matrix $\underset{\sim}{C}$ (that is, the column space of $\underset{\sim}{X}$ is invariant under the matrix $\underset{\sim}{\Omega}$), $\underset{\sim}{t}'\dot{\underset{\sim}{\sigma}}^2$ is not only the b.l.u.e. of $\underset{\sim}{t}'\underset{\sim}{\sigma}^2$ under the derived model free of assumptions concerning $\underset{\sim}{\Omega}$, but also, it is the uniformly best quadratic unbiased estimator (B.Q.U.E.) of $\underset{\sim}{t}'\underset{\sim}{\sigma}^2$. R.D. Anderson, H.V. Henderson, Pukelsheim and Searle (in preparation) appeal to (29) in providing a general proof, free of assumptions concerning $\underset{\sim}{V}$ and the kurtosis parameters, that MINQUE (and REML ignoring non-negativity) are B.Q.U.E. (that is, identical to ANOVA estimators) for balanced data. That special case aside, whenever the pre-assigned variance-covariance matrix $\underset{\sim}{V}_o$ actually equals $\underset{\sim}{V}$ and the kurtosis parameters are, in fact null, $\underset{\sim}{t}'\dot{\underset{\sim}{\sigma}}^2$ is the MIVQUE of $\underset{\sim}{t}'\underset{\sim}{\sigma}^2$. (For $\underset{\sim}{V}_o = \underset{\sim}{V}$ and zero kurtosis, it is easily shown that $\underset{\sim}{r}(\underset{\sim}{X}'\underset{\sim}{\Omega}_o^{-}\underset{\sim}{X}) = \underset{\sim}{r}(\underset{\sim}{X})$ and

$\underset{\sim}{X}'\underset{\sim}{\Omega}_o^-\underset{\sim}{\Omega}(\underset{\sim}{I} - \underset{\sim}{X}(\underset{\sim}{X}'\underset{\sim}{X})^-\underset{\sim}{X}') = \underset{\sim}{0}$, which following C.R. Rao (1971c) are the pertinent necessary and sufficient conditions for MINQUE of $\underset{\sim}{t}'\underset{\sim}{\sigma}^2$ to be MIVQUE.) Thus, if the vector $\underset{\sim}{u}$ is normally distributed, $\underset{\sim}{t}'\dot{\underset{\sim}{\sigma}}^2$ is a MIVQUE if and only if the *a priori* vector of variance parameters $\underset{\sim}{\sigma}_o^2$ is identically equal to the unknown vector $\underset{\sim}{\sigma}^2$ to be estimated (unless, of course, the special case defined by (29) is satisfied). Considering that kurtosis parameters are rarely known, if at all; that when invoked, an assumption of normality is seldom unequivocally true; and the low probability that $\underset{\sim}{\sigma}_o^2 = \underset{\sim}{\sigma}^2$, then, the special case of balanced data aside (29), how strong are the minimum sampling variance properties of MINQUE and MIVQUE of variance parameters likely to be in practice? (Note that as the probability of $\underset{\sim}{\sigma}_o^2 = \underset{\sim}{\sigma}^2$ increases, the need to estimate $\underset{\sim}{\sigma}^2$, at all, diminishes.)

The form of the MINQUE equations (28) suggests an iterative scheme until a "stable" solution for $\underset{\sim}{\sigma}^2$ is obtained. This procedure is known as iterative- or I-MINQUE and although the dependence of the final solution on the *a priori* vector $\underset{\sim}{\sigma}_o^2$ is expected to be reduced using the procedure, it should be noted that, as a consequence of the iteration, the unbiasedness and minimum variance properties of MINQUE and MIVQUE will not be retained. Assuming normality, together with the requirement for non-negative estimates of variance parameters, I-MINQUE solutions to equations (28) are identically REML solutions using the Fisher-scoring computing algorithm (Patterson and R. Thompson, 1974). With REML, the likelihood function associated with a vector of g linearly independent contrasts, $\underset{\sim}{K}\underset{\sim}{y}$, rather than the full likelihood, is maximized to generate equations in the unknown variance parameters. Through the vector $\underset{\sim}{K}\underset{\sim}{y} * \underset{\sim}{K}\underset{\sim}{y}$ being a Kronecker product of the vector of contrasts $\underset{\sim}{K}\underset{\sim}{y}$ with itself, the MINQUE equations (28) were derived following minimization within a restricted sample space identical to that of REML.

If one were to apply ordinary, rather than generalized, least squares principles to the derived model (22) (equivalently, taking *a priori* all σ_i^2, except σ_e^2, equal to zero and all kurtosis parameters to be zero, the MINQUE-0 equations (see Searle, 1979) introduced by Seely (1971) and Hartley, J.N.K. Rao and La Motte (1978) for estimating variance parameters would result.

IV ML-Related Quadratic Estimators

Once again, estimators of $\underset{\sim}{t}'\underset{\sim}{\sigma}^2$ of the form $y'\underset{\sim}{A}y$ will be derived, but in this case, the minimization will not be performed within a restricted sample space of the vector $\underset{\sim}{y}$.

Assume, for the moment, that the vector of fixed effects in the mixed model (1), $\underset{\sim}{\beta}$, is known and consider the quadratic form $(\underset{\sim}{y} - \underset{\sim}{X}\underset{\sim}{\beta})'\underset{\sim}{A}(\underset{\sim}{y} - \underset{\sim}{X}\underset{\sim}{\beta})$ for some symmetric matrix $\underset{\sim}{A}$. As in the previous section, the matrix $\underset{\sim}{A}$ is to be determined such that the variance of the estimator of $\underset{\sim}{t}'\underset{\sim}{\sigma}^2$ will be minimized. In the manner of the previous section,

$$(y - X\beta)'A(y - X\beta) = (\text{vec}A)'[(y - X\beta)*(y - X\beta)], \tag{30}$$

with expected value

$$E[(y - X\beta)'A(y - X\beta)] = (\text{vec}A)'\Psi\sigma^2 , \tag{31}$$

and variance

$$\text{var}[(y - X\beta)'A(y - X\beta)] = (\text{vec}A)'F(\text{vec}A) . \tag{32}$$

That is, the vector $(y - X\beta)*(y - X\beta)$ has mean vector $\Psi\sigma^2$ and variance-covariance matrix F so that the derived model in this instance is

$$(y - X\beta)*(y - X\beta) \sim (\Psi\sigma^2, F) . \tag{33}$$

Once again, there is dependence between the parameter vector σ^2 and the variance-covariance matrix F in the derived model, necessitating *a priori* assumptions. Assigning V_o to V and setting the kurtosis parameters to zero, as before, the "known" variance-covariance matrix for the derived model is

$$F_o = (V_o * V_o)(I + I_{(N,N)}) . \tag{34}$$

A generalized inverse of F_o is $F_o^- = \frac{1}{4}(I + I_{(N,N,)})(V_o^{-1} * V_o^{-1})$, (35)

and it is easily shown that $F_o(F_o)^-\Psi = \Psi$, so that the b.l.u.e. of $t'\sigma^2$, under the derived model with β assumed known and F_o in place of F, is $t'\dot{\ddot{\sigma}}^2$, where $\dot{\ddot{\sigma}}^2$ is a solution to the normal equations

$$(\Psi' F_o^- \Psi)\dot{\ddot{\sigma}}^2 = \Psi' F_o^- [(y - X\beta)*(y - X\beta)] .$$

These equations reduce to

$$\{\mathrm{tr}(V_o^{-1}V_iV_o^{-1}V_j)\}\dot{\sigma}^2 = \{(y - X\beta)'V_o^{-1}V_iV_o^{-1}(y - X\beta)\},$$
$$i, j=1, \ldots, k \ . \qquad (36)$$

In any situation in which β was known, V_o was the actual value of V and the kurtosis parameters were, in fact, zero, $t'\dot{\sigma}^2$ would be the MIVQUE of $t'\sigma^2$. If, further, $F\Psi = \Psi C$ for some matrix C, then $t'\dot{\sigma}^2$ would be the uniformly best quadratic unbiased estimator of $t'\sigma^2$.

Usually, of course, the vector β is unknown and a solution for it is required. Reverting to the original mixed model (1), the b.l.u.e. of an estimable function of β, say $h'\beta$, for given V_o, is $h'\beta^o$ where β^o is a solution to the normal equations $(X'V_o^{-1}X)\beta^o = X'V_o^{-1}y$, that is,

$$\beta^o = (X'V_o^{-1}X)^-X'V_o^{-1}y \ . \qquad (37)$$

Now, using (37) in place of β in (36), the generalized equations for estimating $t'\sigma^2$ become

$$\{\mathrm{tr}(V_o^{-1}V_iV_o^{-1}V_j)\}\tilde{\sigma}^2 = \{y'P_oV_iP_oy\}, \ i, j=1, \ldots, k, \qquad (38)$$

where $\tilde{\sigma}^2$ is to be distinguished from $\dot{\sigma}^2$ (the solution for σ^2 with β known) and P_o is the matrix defined in (26). Equations (38) generally yield biased estimates of $t'\sigma^2$, the reason being that whereas the right-hand-side reflects estimation of the fixed effects in the mixed model, the left-hand-side does not. With P and V being the general representations of P_o and V_o, respectively, it will be observed that the expected value of the ith quadratic form in the right-hand-side of (38) is $E(y'PV_iPy) = \mathrm{tr}(PV_i) \neq \mathrm{tr}(V^{-1}V_i)$ (unless $X = 0$), where $\mathrm{t\tilde{r}}(V^{-1}V_i) = \mathrm{tr}(V^{-1}V_iV^{-1}V_j)\sigma^2$ is the corresponding i^{th} equation in the left-hand side of equations (38). Noting that $\mathrm{tr}(PV_i) = \mathrm{tr}(PV_iPV_j)\sigma^2$ it is immediately apparent (see 28) that MINQUE are unbiased. For completeness, it is noted that $\mathrm{tr}(V^{-1}V_i) = E\,[(y - X\beta)'V^{-1}V_iV^{-1}(y - X\beta)]$ thereby verifying the earlier claim that for known β, equations (38) generate an unbiased estimate of $t'\sigma^2$.

Although equations (38) lead, in general, to biased estimates of variance parameters, there is some appeal in choosing to use those equations ahead of, for instance, the MINQUE counterpart (equations (28)); computing matrix functions involving P_o is generally more difficult than is the case with functions involving V_o^{-1} and; parameters such as heritability are most often estimated by taking a ratio of estimates of variance parameters and irrespective of whether the latter are unbiased, the ratio estimate will usually be a biased estimate.

The form of equations (38) suggests an iterative analogue of I-MINQUE. Assuming normality and invoking the requirement that estimates of variance parameters be non-negative, iterative solutions to equations (38) are identically the ML solutions generated by the Fisher-scoring computing algorithm proposed by T.W. Anderson (1973). This is a rather important result; equations leading to the ML estimate of $t'\sigma^2$ were obtained through the application of G.L.S. principles to a derived model and thus, the dispersion-mean model approach of Pukelsheim (1976) has been extended to provide a unifying theory for the derivation of REML, MINQUE, MIVQUE, I-MINQUE and ML estimators of variance parameters. Moreover, the fact that the derivation theory is based on linear estimation principles is advantageous for (in particular) animal breeders.

Applying ordinary least squares to the derived model (33), together with the replacement of β by $\beta^o = (X'X)^- X'y$, leads to the equations

$$\{tr(V_iV_j)\}\sigma^2 = \{y'MV_iMy\}, \; i, \; j=1, \; \ldots, \; k, \tag{39}$$

for M of (15). Clearly (39) is a special case of (38) and even though (39) furnishes biased estimators, the computational simplicity involved is an appealing feature.

V Resumé of Sections II to IV

Linear estimation principles have been used in a unified approach to the derivations of recently-proposed methods for estimating variance parameters in mixed models. The methods can be broadly divided into two categories; (i) Methods whereby the loss in degrees of freedom associated with the estimation of fixed effects in the model is taken into account (REML, MINQUE, MIVQUE, I-MINQUE, MINQUE-0) and (ii) Methods not taking account of the loss in degrees of freedom associated with estimation of fixed effects (ML, Equations (38), Iteration of Equations (38), Equations (39)).

VI Considerations for the Future

As far as on-going animal breeding programmes are concerned, the mixed model defined in Section II of this paper is too simplistic. The assumptions var($\underset{\sim}{u}_i$) = $\underset{\sim}{I}\sigma_i^2$, i=1, ..., k, and cov($\underset{\sim}{u}_i$, $\underset{\sim}{u}_{i'}$) = $\underset{\sim}{0}$ i≠1' mean that the model will not handle important situations such as:

(1) The use of certain animal relationships (if $\underset{\sim}{u}_i$ is a vector of sire effects, the use of relationships amongst sires leads to var($\underset{\sim}{u}_i$) being non-diagonal; and

(2) the observation vector $\underset{\sim}{y}$ containing, in the multiple-trait case, several measurements on a single animal.

Furthermore, the possibility of selection having previously occurred is ignored. Animal breeders are frequently confronted with the estimation of variance and covariance parameters in mixed models with unbalanced data involving:

(1) Information on more than one trait, often leading to a mixture of continuous and categorical variables with unknown respective distributions and differing mean vectors;

(2) on-going selection and usage of related animals; and

(3) relatively small numbers of data points in the sub-classes defined by combinations of fixed effects in the model.

The development of techniques for estimating variance parameters for categorical traits is a high priority research area. Moreover, estimating parameters free of the effects of selection is a major concern for animal breeders. Whether or not the dispersion-mean concept exhibited in this paper can be exploited for solving some of these problems is unknown at this stage. In view of the fact that genetic parameter estimation is frequently a prelude to the ranking of candidates for selection, there is a strong case for the development of robust ranking procedures rather than emphasizing the usual optimality properties of point estimates of parameters.

ACKNOWLEDGMENTS

The helpful influence of C.R. Henderson, A.L. Rae, S.R. Searle and R.L. Quaas on the author's understanding of parameter estimation is gratefully acknowledged. The author is also indebted to the organizers of this Symposium for the opportunity to participate.

REFERENCES

Anderson, R.D., Quaas, R.L., and Searle, S.R. (1977). *Paper No. BU-630-M, Biometrics Unit, Cornell University, Ithaca, New York.* 11 pp.

Anderson, T.W. (1973). *Ann. Stat. 1,* 135.

Crump, S.L. (1947). Ph.D. Thesis, Iowa State University, Ames.

Hartley, H.O., and Rao, J.N.K. (1967). *Biometrika 54,* 93.

Hartley, H.O., Rao, J.N.K., and La Motte, L.R. (1978). *Biometrics 34,* 233.

Henderson, C.R. (1953). *Biometrics 9,* 226.

Henderson, H.V., and Searle, S.R. (1979). *Canad. J. Stat. 7,* 65.

Kendall, M.G. (1946). "The Advanced Theory of Statistics," Vol. II. Charles Griffen and Co. Ltd., London.

Kleffe, J. (1976). *Math. Oper. & Stat. 7,* 707.

La Motte, L.R. (1970). *Tech. Report, Dept Statistics, University of Kentucky, Lexington, Kentucky.* 13 pp.

Neudecker, H. (1969). *J. Am. statist. Ass. 64,* 953.

Patterson, H.D., and Thompson, R. (1971). *Biometrika 58,* 545.

Patterson, H.D., and Thompson, R. (1974). *Proc. 8th Int. Biometrics Conf.* p. 197.

Pukelsheim, F. (1976). *J. Multivariate Anal. 6,* 626.

Pukelsheim, F. (1977). *Math. Oper. & Stat. 8,* 323.

Rao, C.R. (1970). *J. Am. statist. Ass. 65,* 161.

Rao, C.R. (1971a). *J. Multivariate Anal. 1,* 257.

Rao, C.R. (1971b). *J. Multivariate Anal. 1,* 445.

Rao, C.R. (1971c). *Sankhȳa Ser. A. 33,* 371.

Rao, C.R. (1972). *J. Am. statist. Ass. 67,* 112.

Searle, S.R. (1979). *In* "Variance Components and Animal Breeding" Proc. of a Conf. in Honor of C.R. Henderson, July 16-17, 1979, Cornell University, Ithaca, New York. p. 59.

Seely, J. (1971). *Ann. math. Statist. 42,* 710.

Thompson, W.A. (1962). *Ann. math. Statist. 33,* 273.

Townsend, E.C. (1968). Ph.D. Thesis, Cornell University, Ithaca, New York.
Townsend, E.C., and Searle, S.R. (1971). *Biometrics 27*, 643.
Zyskind, G., and Martin, F.B. (1969). *SIAM J. Appl. Math. 17*, 1190.

Developments in the Use of BLUP for Estimation of Genetic Merit

A. E. McClintock[1]
J. F. Taylor[2,3]

Animal Genetics and Breeding Unit[4]
University of New England
Armidale, N.S.W.

Best Linear Unbiased Prediction methods (BLUP) are being used in a number of livestock species to extract estimates of genetic merit of breeding animals from field performance records. This paper outlines some of the problems associated with the application of BLUP procedures and highlights areas where further developments could take place. Matters of practical importance which are also discussed include connectedness of data, speed of processing, the methods of expressing estimates of breeding merit and their reliability.

I Introduction

The development of the general theory of Best Linear Unbiased Prediction (BLUP) and the availability of high speed computers suited for its application to large data sets has allowed the adoption of the technique for research in many diverse aspects of animal breeding. Schaeffer (1979) has

[1]*Supported by the Australian Meat Research Committee.*
[2]*Present address: Cornell University, Ithaca, New York.*
[3]*Supported by the Australian Dairying Research Committee.*
[4]*A joint Unit of the New South Wales Department of Agriculture and the University of New England.*

FUTURE DEVELOPMENTS IN THE GENETIC IMPROVEMENT OF ANIMALS
ISBN 0 12 078830 6

recently drawn together many of the concepts developed by Henderson and his Cornell co-workers, but for those wishing to study an early authoritative work, Henderson (1975) is a key reference.

While the initial proponents for the use of the theory in animal breeding were primarily concerned with dairy cattle populations, where it has been extensively applied, the theory and techniques for efficient application are now evident in other species. Hallgrimsson and Jonmundson (1979) have predicted breeding values of Icelandic sheep for carcass weight, Van Vleck and Hintz (1975) predicted the genetic value of stallions for race performance, Van Vleck (1976), Eriksson, Wilton and Henningsson (1979), Slanger (1979, 1980) and Quaas and Pollak (1980) derive and apply operational models to predict genetic differences among beef sires for traits including weaning weight, post-weaning gain and yearling weight. The model employed by Quaas and Pollak (1980) illustrates the flexibility of BLUP application in that the weaning weight of a calf is partitioned to include the contributions of sire's genotype for growth and dam's genetic and permanent environmental components of maternal ability.

Our attempt to highlight areas for future development of BLUP applications and theory may appear to be a disjointed list of seemingly unrelated topics. The common thread, of course, is the computation and presentation of Best Linear Unbiased Predictors using field data. Most of the subjects highlighted may be viewed as problem areas which require further research and discussion.

II Sire Groups in Relation to Connectedness

One of the features of early BLUP approaches to sire evaluation involved grouping of sires with a similar genetic background and solving for groups as though they were fixed effects. Usually groups will coincide with annual cohorts of test bulls and frequently bulls belonging to different AI organizations will be put into separate groups.

If two groups are only weakly connected, serious biases can occur. To illustrate the problem, consider an extreme situation. Suppose it is desired to compare test bulls from two different states (or countries) but that there are very few within-herd comparisons of bulls represented in both programs. By chance, atypical daughter record comparisons of these cross-linking sires could result in a large (and unreliable) difference between the mean BLUPs of bulls in both programs. However, if bulls in both programs were placed in the same group then this unreliable contrast would not have such a pronounced effect, especially when the sires had relatively few progeny.

As a general policy, bulls should not be subdivided into groups unless:

(1) There is a large number of within-herd cross-links between bulls in these groups (see Section III).
(2) There are reasonable *a priori* grounds for believing that the creation of groups is necessary (i.e., the groups of bulls are from genetically distinct sub-populations).

Grouping all bulls entering AI in a particular year (without subdividing them according to testing program to which they belong) would avoid some of the problems associated with poor connectedness. This would bias evaluations if a genetic difference existed between testing programs, but would seem to be a practical and conservative approach. Deliberate steps should also be taken to improve the pattern of bull usage as soon as practicable, thereby avoiding such problems.

Perhaps some form of adjustment could be devised to draw together (regress) the solutions of weakly connected groups. For example, groups could be treated as a random effect. This would involve adding the ratio of error to group variances to the diagonal coefficients of the group x group equations (in many cases this ratio would be in the order of 10^3). Such a pragmatic solution would not appeal to a purist.

In circumstances where all genetic relationships between sires are included in the inverse numerator relationship matrix (NRM), the need for grouping is largely avoided.

III Connectedness

The concept of connectedness can play an important role in the analysis of production records. When estimated breeding values are used to select superior animals, invalid comparisons may be made if the data are not connected. The need for connectedness is only of importance when the mean genetic values of disconnected groups differ, as these differences are not estimable. When data are extensive, disconnected groups may go undetected and if trend or strain differences exist between groups these will not be accounted for in the model, causing comparisons of estimated breeding values between groups to be invalid.

Connectedness, ensuring valid comparisons, arises when:

(1) Within Herd-Year-Season (HYS) contrasts, or
(2) Genetic relationships between animals exist.

Disconnectedness, preventing valid comparisons, is caused by separation of progeny groups in:

(1) Time, or
(2) Space (geography).

Connectedness can be either direct or indirect; direct if two sires have progeny in the same HYS; indirect if two sires with progeny in different HYSs are each directly linked to a common third sire. Genetic relationships are incorporated in BLUP by using the Numerator Relationship Matrix (NRM) and are always direct; two sires are either related genetically or they are not.

In practice the degree of connectedness is not calculated, although such a measure could help highlight weak connections without the need to invert a large sire-by-sire matrix of coefficients. Genetic relationships can exist between individual sires (e.g., maternal half-sibs) but, because of simplifying assumptions, these connections are not represented in the NRM.

In all but the most trivial of designs it will not be possible to ascertain connectedness visually. However, for analyses with a limited number of sires represented, a test for connectedness which is computationally inexpensive was illustrated by Petersen (1978). When BLUP analyses are conducted, solutions for herd (herd-year or herd-year-season) effects are not usually required and herd equations are absorbed into equations for sires as the mixed model equations are set up in computer core. The structure of the sire x sire portion of these equations, with herds absorbed, is such that a non-zero element on the $(i,j)^{th}$ off-diagonal position only occurs if sires i and j have progeny in the same herd. Sweeping rows of this matrix provides a list of connected groups, and detects both direct and indirect (within HYS) connections. Similarly, the NRM or inverse NRM may be swept to determine genetically connected groups. However, as the inverse NRM is added to the sire x sire portion of the mixed model equations prior to solving to obtain BLUPs of breeding values, the test may be left until after addition of the inverse NRM to reveal connectedness of either type. This test has been programmed in FORTRAN, sweeps out the elements to the right of the diagonal of a full-stored matrix, outputting connected groups and is available from the authors. Unfortunately, scale is a major limitation, depending on the magnitude of the matrix to be stored in core. Further, this test is unable to determine the strength or number of links existing between subgroups which have been found to form a single connected group.

IV Genetic Bases—Fixed or Rolling?

In most mixed linear models currently used for bull and cow evaluation absolute breeding values cannot be estimated, but differences between any two breeding values can be uniquely determined. This is because only deviations from group means and differences between groups can be estimated. Genetic groupings of sires or cows are used to differentiate between genetically distinct animals in time, thus representing genetic trend. Two alternatives for estimating group means are to set one group (year) to zero or let the sum of group solutions be zero. The first alternative represents a Fixed Base to which all subsequent groups are compared, while the second represents a Rolling Base which changes as new data are included in the analysis. There are variations on the rolling base principle, for example:

(1) The latest group (year) can always be set to zero (as is done, for example, in Norway).

(2) The weighted average of all sire BLUPs can be set to zero; the weightings being the number of progeny in a particular (rolling) time period.

(3) The base could be repositioned after a fixed time period (Dickinson, 1973 termed this a 'stepwise base').

The rankings of sires evaluated using a BLUP method will not be affected by choice of genetic base. The decision of what base to use is both a marketing and a philosophical problem - marketing, in that most owners would prefer not to have negative proofs for their bulls - philosophical, in that some people prefer a rolling base while others prefer a fixed base.

The main advantage of the rolling base is that many people have become used to the idea through use of contemporary comparisons. However, as advocates of the fixed base, we regard the rolling base of the contemporary comparison to be its major disadvantage. Other disadvantages of the rolling base are:

(1) All documentation (for example, sire pedigrees showing EBVs of ancestors) will become outdated once the next analysis is completed.

(2) Users of the BLUP results do not have the psychological stimulation of observing genetic progress; some bulls in the next listing will be better than those currently available.

(3) If the latest cohort of bulls to be evaluated is poorer than the previous cohort a rolling base often tends to obscure this fact.

(4) It is more difficult to explain annual reports to farmers which show the genetic and environmental trends within a herd.

(5) Sire evaluations in Canada, New Zealand and U.K. employ fixed bases which give AB organizations in these countries a marketing advantage when competing with AB organizations whose bulls have been compared with a rolling base. It is likely that bulls will be more saleable when their Estimated Breeding Values are large positive figures.

(6) If genetic bases in different countries are constantly changing with respect to each other, international comparison of sires is made even more difficult.

V Methods of Expressing Genetic Merit and the Amount of Information in an Estimate

There are four possibilities for expressing genetic merit, viz:

(1) Individual's own estimated genetic merit expressed as a deviation in kg, metres, litres per minute, etc. This is usually referred to as a Breeding Value (BV) or more accurately, Estimated Breeding Value (EBV).

(2) Individual's own estimated genetic merit measured as a percentage of the phenotypic population mean. This is usually referred to as Relative Breeding Value (RBV).

(3) Analagous to (1) but the deviations are all halved to indicate the expected genetic contribution to the progeny. Estimated Transmitting Ability is the usual nomenclature (ETA).

(4) Analagous to (2) but the percentages are all scaled to indicate the expected progeny merit. This method is used in some European countries and could be called a Relative Transmitting Ability (RTA). Thus an RBV of 110 is equivalent to an RTA of 105.

Although RBVs or RTAs are probably more meaningful over a range of levels of production, the use of actual units (e.g., kg) is more acceptable to breeders and less likely to be confused with measures of reliability or accuracy of tests. Progeny test results naturally lend themselves to being expressed as ETAs or RTAs. However, when a measurement on an individual's own performance is also used (e.g., a cow's milk production, or a bull's growth rate) confusion can be caused unless a convention is adhered to. Consider the following expected relationships:

$$EBV_{progeny} = [EBV_{sire} + EBV_{dam}] / 2 \tag{1}$$

$$EBV_{progeny} = ETA_{sire} + ETA_{dam} \tag{2}$$

$$ETA_{progeny} = [ETA_{sire} + ETA_{dam}] / 2 \tag{3}$$

Equation 2 is to be avoided since it uses both concepts and is therefore unnecessarily complex. The concept of $ETA_{progeny}$ is not easily grasped since it involves an additional generation.

For these reasons, we prefer the use of EBVs to the exclusion of the other methods of expression. Sire BLUPs based on progeny information therefore need to be doubled before being published as EBVs.

There are a number of ways of indicating how much reliance can be placed on the BLUP for an individual animal.

(1) Number of daughters.

(2) Effective number of daughters.

(3) Standard Error of Prediction (SEP) or Prediction Error Variance (PEV).

(4) A star system based on any of the above.

Sadly, none of these methods in practice solves the primary problem which is that of potential bias. Terms such as accuracy, reliability and repeatability of the selection index are commonly used, but the usual method for estimating the accuracy of an index presupposes that no biases exist. Perhaps "precision" would be a more meaningful term, especially to statisticians. In the absence of bias, a BLUP for an individual has an equal probability of over- or under-estimating the individual's true merit, so selection of superior individuals should be done on the basis of their BLUPs alone.

What users are particularly fearful of, is the deliberate biasing of a bull's EBV. In cases where progeny appear only in one or two herds, perhaps in the same ownership as the bull, this fear is a reasonable one and the risk cannot be quantified. The publication of a bull's daughter distribution across herds can help to identify such potential risks.

However, users are likely to want some measure of the stability of a BLUP in cases where only one or two bulls are to be used and where chance could play an important role. In such circumstances, the effective number of daughters is to be preferred to actual number for obvious reasons, but neither would indicate the impact of the NRM on a sire's BLUP.

The use of Prediction Error Variances (PEV) in, for example, kg^2 units seems unlikely to be acceptable to users. The use of $PEV^{\frac{1}{2}}$ or SEP is being used in, for example, the Canadian Direct Sire Comparison system. It requires care to explain the meaning of SEPs (or Confidence Intervals) and in some circumstances could discredit the breeding program where SEPs are large relative to between-bull differences. SEPs for cow evaluations would be very large. Users might be tempted to demand very small SEPs or might select on other traits altogether (e.g., colour of tails!). In addition, some sires' predictions will change by more than twice their SEPs which may make users distrust the method.

If EBVs are used to express genetic merit then, to avoid confusion, the measure of accuracy of the BLUP might be expressed as a ratio (i.e., a unitless measure). We consider various alternatives below.

One could express the information as a ratio (A) of PEV to the variance component due to sires (i.e., how large is the PEV relative to the between-bull variance?) thus:

$$A = (PEV_w) / \sigma_s^2$$

where σ_s^2 is the variance due to sires and PEV_w is the Prediction Error Variance for a sire with w effective daughters.

It can be shown that for the ith sire

$$A = C_{ii} \lambda$$

where C_{ii} is the ith diagonal of the inverse from the sire x sire portion of the LHS coefficient matrix,

$$\lambda = (4/h^2) - 1$$

and h^2 is the heritability of the trait. Since C_{ii} is usually expensive to compute, $1/(w + \lambda)$ is often taken as an approximation. In these circumstances

$$A \approx \lambda/(w + \lambda) = 1 - w/(w + \lambda)$$

A will vary between 0 and 1, $A = 0$ as $w \rightarrow \infty$. Define a new ratio A' as follows

$$A' = 1 - A \approx w/(w + \lambda)$$

Note that $A' = 1$ as $w \rightarrow \infty$ and that the correlation (r_{TI}) between True Breeding Value and the BLUP is:

$$r_{TI} = A'^{\frac{1}{2}} \approx [wh^2 \ / \ (4 + (w - 1)h^2)]^{\frac{1}{2}} = [w/(w + \lambda)]^{\frac{1}{2}}$$

Geneticists would probably accept r_{TI} as a measure of accuracy (precision) but it is debatable whether A' is more meaningful than $A'^{\frac{1}{2}}$.

These ratios are frequently multiplied by 100 to convert them to percentages.

Finally, a star system based on any of the above criteria suffers from the discrete nature of the scale. Usually, sires are graded *, ** or ***, which may be too crude a distinction for a user who wishes to distinguish between two bulls having almost identical accuracies but different star grades. In addition, separate star systems would be needed for bulls and for cows; this could lead to confusion.

We favour the use of r_{TI} expressed as a percentage.

VI Environmental Correlations (c^2)

Van Vleck (1966a) concluded that there was an environmental correlation (c^2) between paternal half-sibs in the same herd-year, which was as large as the expected correlation between paternal half-sibs (0.08) for Holstein first lactation milk yield. However, Van Vleck (1966b) concluded that because of the pattern of daughter distribution for AI sires, this was of no practical importance. The subsequent introduction of liquid Nitrogen frozen semen may have now caused the c^2 effect to become important even for AI sires.

Henderson (1973) has shown how c^2 may be incorporated in BLUP. However its incorporation greatly complicates the procedure for estimating breeding values and the extra cost may not justify the increased precision, even if feasible.

VII Categorical Trait Analysis

There are three approaches being used to analyse data arising from categorically scored traits at present:

(1) To assume an underlying continuous normal distribution and to apply a score to each class to ensure that a normal distribution is produced (Schaeffer and Burnside, 1974; Schaeffer and Wilton, 1976; Schaeffer, Hunt and Burnside, 1978; Berger and Freeman, 1978).

(2) To treat each category as a 0-1 score and fit a multiple trait BLUP model with each category being treated as a separate trait with known covariance structure (Van Vleck, 1980; Van Vleck and Karner, 1980; Quaas and Van Vleck, 1980).

(3) To treat categories as 0-1 but assume the model for each category to be multiplicative (contingency tables) and not linear (Gianola, 1980a, b).

There is a need to decide which of these methods is best suited to each problem; which mathematical model is most appropriate to the biology of the situation. In addition, robustness and ease of computation should obviously be considered when distinguishing between methods. Another unresolved problem is how to include fixed effects in the model; the (co)variance structure between categories will vary within levels of these fixed effects.

VIII Economic Evaluation

There is a close parallel between linear Selection Indices (Hazel, 1943) and BLUP. By applying economic weights to Estimated Breeding Values obtained from a multiple trait BLUP model, an overall economic evaluation can be made for each individual. The economic weights will be identical to those required for a Selection Index calculation.

There seems to be no theory to allow the BLUP equivalent of restricted Selection Indices to be calculated. For traits having intermediate optima, a quadratic term could be included in the multiple regression equation used to define total economic merit.

IX Preadjustment of Data

Historically, the only practical way of removing systematic environmental effects has been to preadjust the data before calculating Breeding Values.

The philosophical argument for preadjusting data prior to analysis rather than include effects in the model relies upon the assumption that solutions for these effects remain invariant in time and space, provided estimates have been obtained from application of the true biological model to a large data set which is representative of the population. For example, under these assumptions Everett, Taylor and Hammond (1982) found that depending on breed, a cow calving at 20 months of age in an average NSW herd-year was estimated to produce 50 to 75 percent of another cow of the same breed calving at 84 months of age in an equivalent herd-year.

Provided the operational model is in fact the true biological model, preadjustment of data will not bias estimated genetic differences between cows. However, many of the difficulties encountered in the analysis of large data sets collected in the field invariably place limitations on the sophistication of operational models and invalidate this assumption. For example, the cross-classified operational model employed by Everett *et al.* (1982) would imply that additive adjustment factors are biologically appropriate. However, cows calving at 20 months of age in a high producing herd will produce relatively more at 84 months of age than a similar cow in a low producing herd (Miller, 1973). To alleviate this problem, Everett *et al.* (1982) converted estimated additive adjustment factors into multiplicative factors which are implied by the percentage figures given above. While this approach may appear to be dubious, an alternative is to calculate additive adjustment factors within classifications of mean herd levels of production if this is computationally feasible. Significant problems are likely to arise if bulls' daughters are poorly distributed over age at calving classifications and inappropriate adjustment factors are used.

Another problem inherent in preadjustment of data which is not generally recognized by practitioners is that the distributional properties of solutions for effects remaining in the model after adjustment are altered. When the true biological model has been used to obtain additive adjustment factors, it can be shown that while subsequent analyses employing these factors produce unbiased solutions for the reduced system of equations, Sampling and Prediction Error Variances (SPEVs) are larger than those calculated assuming known adjustment factors. However, because sampling variances of additive adjustment factors are likely to be small if they have been estimated from large data sets, SPEVs of solutions to the reduced model after adjustment should be smaller than those arising from a model in which all effects are estimated simultaneously. When an incorrect operational model is assumed or multiplicative adjustment factors used, solutions to the reduced system may also be biased.

X Use of Estimated Population (Co)Variances

Henderson's (1973) derivation of BLUP of effects in a mixed linear model is analagous to the derivation of weighted least squares (WLS) estimators in a fixed effects model. Consequently, the only assumption required for BLUP is that (co)variances describing random genetic and environmental effects in the model are known. In practice, heritabilities, repeatabilities and environmental and genetic (co)variances are never known, although 'good' estimates of these parameters may exist. If this is the case, the usual procedure is then to substitute estimated components for the unknown (co)variance parameters in the model regardless of the consequences of this action.

The purpose of this section is to provide the distributional properties of solutions to Henderson's (1973) mixed model equations when (co)variance components estimated independently of the current data for analysis, $\underset{\sim}{y}$, replace the true but unknown parameters in the model.

If we write the general mixed linear model as:

$$\underset{\sim}{y} = X\underset{\sim}{\beta} + Z\underset{\sim}{u} + \underset{\sim}{e}$$

with

$$E\begin{bmatrix}\underset{\sim}{y}\\ \underset{\sim}{u}\\ \underset{\sim}{e}\end{bmatrix} = \begin{bmatrix}X\underset{\sim}{\beta}\\ \underset{\sim}{0}\\ \underset{\sim}{0}\end{bmatrix} \qquad \mathrm{Var}\begin{bmatrix}\underset{\sim}{y}\\ \underset{\sim}{u}\\ \underset{\sim}{e}\end{bmatrix} = \begin{bmatrix}ZGZ' + R & ZG & R\\ & G & 0\\ \text{sym.} & & R\end{bmatrix}$$

then Henderson (1973) has shown that the BLUP of the estimable linear contrast $\underset{\sim}{k}'\underset{\sim}{\beta} + \underset{\sim}{m}'\underset{\sim}{u}$ is $\underset{\sim}{k}'\underset{\sim}{\hat{\beta}} + \underset{\sim}{m}'\underset{\sim}{\hat{u}}$, where $\underset{\sim}{\hat{\beta}}$ and $\underset{\sim}{\hat{u}}$ are solutions to:

$$\begin{bmatrix}X'R^{-1}X & X'R^{-1}Z\\ Z'R^{-1}X & Z'R^{-1}Z + G^{-1}\end{bmatrix}\begin{bmatrix}\underset{\sim}{\hat{\beta}}\\ \underset{\sim}{\hat{u}}\end{bmatrix} = \begin{bmatrix}X'R^{-1}\underset{\sim}{y}\\ Z'R^{-1}\underset{\sim}{y}\end{bmatrix} \qquad (4)$$

If $\begin{bmatrix}C_{11} & C_{12}\\ C_{12}' & C_{22}\end{bmatrix}$ is the inverse of the coefficient

matrix in (4) after constraint to full rank we have:

$$\mathrm{Var}\begin{bmatrix}\underset{\sim}{\hat{\beta}}\\ \underset{\sim}{\hat{u}}\end{bmatrix} = \begin{bmatrix}C_{11} & 0\\ 0 & G - C_{22}\end{bmatrix}; \quad \mathrm{Var}\begin{bmatrix}\underset{\sim}{\hat{\beta}}\\ \underset{\sim}{\hat{u}} - \underset{\sim}{u}\end{bmatrix} = \begin{bmatrix}C_{11} & C_{12}\\ C_{12}' & C_{22}\end{bmatrix} \qquad (5)$$

Now suppose that G and R are unknown although estimates A and S have been made independently of the current data for analysis, $\underset{\sim}{y}$. Then it can be shown that:

$$E\left[\begin{matrix}\underset{\sim}{\hat{\beta}} - \underset{\sim}{\beta}\\ \underset{\sim}{\hat{u}} - \underset{\sim}{u}\end{matrix}\,\middle|\, A,S\right] = E\begin{bmatrix}\underset{\sim}{\hat{\beta}} - \underset{\sim}{\beta}\\ \underset{\sim}{\hat{u}} - \underset{\sim}{u}\end{bmatrix} = \begin{bmatrix}\underset{\sim}{0}\\ \underset{\sim}{0}\end{bmatrix}$$

and hence solutions to (4) with estimates A and S replacing matrices G and R respectively, provide conditionally and unconditionally unbiased solutions provided elements of A and S are estimated independently of $\underset{\sim}{y}$.

Further, if we define:

$$N = \begin{bmatrix}X'S^{-1}RS^{-1}X & X'S^{-1}RS^{-1}Z\\ Z'S^{-1}RS^{-1}X & Z'S^{-1}RS^{-1}Z + A^{-1}GA^{-1}\end{bmatrix}$$

and $M = \begin{bmatrix} X'S^{-1}X & X'S^{-1}Z \\ Z'S^{-1}X & Z'S^{-1}Z + A^{-1} \end{bmatrix}^{-1}$

after constraint to full rank, then

$$\mathrm{Var}\left[\begin{array}{c} \hat{\underset{\sim}{\beta}} \\ \hat{\underset{\sim}{u}} - \underset{\sim}{u} \end{array} \middle| A,S\right] = MNM$$

and

$$\begin{aligned} \mathrm{Var}(\hat{\underset{\sim}{u}}|A,S) &= G - C_{22}A^{-1}G - GA^{-1}C_{22} + \mathrm{Var}(\hat{\underset{\sim}{u}} - \underset{\sim}{u}|A,S) \\ \mathrm{Cov}(\hat{\underset{\sim}{\beta}},\underset{\sim}{u}'|A,S) &= -C_{12}A^{-1}G \\ \mathrm{Cov}(\hat{\underset{\sim}{u}},\underset{\sim}{u}'|A,S) &= G - C_{22}A^{-1}G \end{aligned} \qquad (6)$$

When A = G and R = S the unconditional (co)variances derived by Henderson (1973) reported in (5) are obtained. When R = S result (6) simplifies to yield Henderson's (1975) conditional result.

To obtain unconditional Sampling and Prediction Error (Co)variances (SPEVs) from the conditional SPEVs in (6) we have:

$$\mathrm{Var}\left[\begin{array}{c} \hat{\underset{\sim}{\beta}} \\ \hat{\underset{\sim}{u}} - \underset{\sim}{u} \end{array}\right] = E\left[\mathrm{Var}\left[\begin{array}{c} \hat{\underset{\sim}{\beta}} \\ \hat{\underset{\sim}{u}} - \underset{\sim}{u} \end{array} \middle| A,S\right]\right] + \mathrm{Var}\left[E\left[\begin{array}{c} \hat{\underset{\sim}{\beta}} \\ \hat{\underset{\sim}{u}} - \underset{\sim}{u} \end{array} \middle| A,S\right]\right] \qquad (7)$$

$$= E(MNM) \qquad (8)$$

as the second term in (7) is null. The expectation in (8) is made over the sample space of A and S. The unconditional SPEVs derived in (8) account for the additional variability due to incorporating estimated parameters in place of the true but unknown values.

Clearly a closed solution for (8) will not generally be possible due to the complicated matrix function of A and S involved. However, for particular cases, approximate results for unconditional SPEVs may be obtained. Taylor (1982) examined the special case where $R = I\sigma_e^2$, $G = I\sigma_s^2$, $S = I\hat{\sigma}_e^2$ and $A = I\hat{\sigma}_s^2$, where $\hat{\sigma}_e^2$ and $\hat{\sigma}_s^2$ are estimates of σ_e^2 and σ_s^2 respectively. In this case (8) simplifies and Taylor (1982) was able to use partitioned matrix results and a matrix series expansion to approximate (8). Results of a Monte Carlo computer simulation based on the one-way random model indicated that unconditional SPEVs are remarkably robust to estimates of $\lambda = \sigma_e^2/\sigma_s^2 = 4/h^2 - 1$, provided the mean square error of the estimates is 'not too large'.

Example: Heteroscedastic Error Variances Within Herds

Consider the model:

$$y_{ijk} = \mu + hys_i + s_j + e_{ijk}$$

$i = 1,..,p \quad j = 1,..,q$ and $k = 1,..,n$

where y_{ijk} is the kth observation (daughter record) on the jth sire in the ith herd-year-season. We have:

$G = I_q\sigma_g^2$ is a q x q matrix,

$R = \sum_{i=1}^{p}{}^{+} I_{np}\sigma_i^2$ is a npq x npq matrix,

σ_g^2 is the additive genetic variance associated with sires,

σ_i^2 $i = 1,...,p$ are heterogeneous within-herd error variances, and

Σ^+ is the direct sum.

Because μ is not estimable, we constrain the solution $\hat{\mu} = 0$. This has the consequence that $E(\hat{hys}_i) = \mu + hys_i$.

After some algebra it can be shown that the correlation between true breeding value and BLUP for each individual, based on np progeny is:

$$r_{TI} = \left\{\frac{n(q - 1)\varepsilon.\sigma_g^2}{q(n\varepsilon.\sigma_g^2 + \gamma)}\right\}^{\frac{1}{2}} \tag{9}$$

where $\gamma = \prod_{i=1}^{p} \sigma_i^2$

$$\varepsilon. = \sum_{i=1}^{p} \prod_{v \neq i=1}^{p} \sigma_v^2$$

However, σ_g^2 and $\sigma_i^2 = 1,..,p$ are not known and replaced by estimates $\hat{\sigma}_g^2$ and $\hat{\sigma}_e^2$. Hence heterogeneity of error variances is ignored and a single parameter estimate used to replace p unknown parameters. The correlation between true breeding value and BLUP conditional on $\hat{\sigma}_e^2$ and $\hat{\sigma}_g^2$ can be shown to be:

$$r_{TI}|\hat{\sigma}_e^2,\hat{\sigma}_g^2 = \left\{\frac{np^2(q-1)\sigma_g^2}{q(np^2\ \sigma_g^2+\sigma_\cdot^2)}\right\}^{\frac{1}{2}} \tag{10}$$

where $\sigma_\cdot^2 = \sum_{i=1}^{p}\sigma_i^2$

Notice that (10) is independent of $\hat{\sigma}_e^2$ and $\hat{\sigma}_g^2$, this appears to be a consequence of completely balanced designs.

$$\text{Define Ac} = \frac{r_{TI}|\hat{\sigma}_e^2,\hat{\sigma}_g^2}{r_{TI}}$$

$$= \left\{\frac{p^2(n\varepsilon.\sigma_g^2+\gamma)}{\varepsilon.(np^2\sigma_g^2+\sigma_\cdot^2)}\right\}^{\frac{1}{2}} \qquad \text{from (9) and (10)}$$

It is not possible to provide an unconditional value for (7) unless it is known how $\hat{\sigma}_e^2$ and $\hat{\sigma}_g^2$ have been estimated. However, for the conditional case with balanced data,

(1) Ac is independent of $\hat{\sigma}_g^2$ and $\hat{\sigma}_e^2$ and q the number of sires,
(2) Ac $\to$ 1 as $n \to \infty$ for σ_g^2, $\sigma_i^2 < \infty$ $\quad i = 1, \ldots, p$.
(3) Ac = 1 when $\sigma_i^2 = \sigma_e^2$ for all $i = 1,\ldots,p$.

Table I contains values for $r_{TI}, r_{TI}|\hat{\sigma}_e^2,\hat{\sigma}_g^2$ and Ac for a range of values of n, σ_1^2 and σ_2^2 when p = 2, q = 3 and $\sigma_g^2 = 1$.

These results for a completely balanced design are to be expected. The larger the differences between σ_1^2 and σ_2^2, the less efficient will be a model assuming homoscedastic error variances. As this difference becomes larger, the effective number of progeny per sire decreases and the correlation between true breeding value and BLUP must decrease. As σ_i^2 increases, the within-herd heritabilities $h_i^2 = 4/(\sigma_i^2 + 1)$ $i = 1,2$ (as $\sigma_g^2 = 1$) decrease and so do r_{TI} and $r_{TI}|\hat{\sigma}_e^2,\hat{\sigma}_g^2$. Observe that Ac $\to$ 1 as $n \to \infty$. For the completely balanced design with all sires having equal representation in all herds one might expect differences in σ_e^2 between herds to have little effect on the efficiency of selection and this is true unless there are large differences between σ_i^2, heritabilities are low and sires have few daughters. In this latter case no information can afford to be wasted.

TABLE I. Values of r_{TI}, $r_{TI}|\sigma_e^2, \sigma_g^2$ *and Ac for Balanced Designs.*

σ_1^2	σ_2^2	n	r_{TI}	$r_{TI}\|\hat{\sigma}_e^2, \hat{\sigma}_g^2$	Ac
5	10	1	0.39	0.37	0.96
		10	0.71	0.70	0.98
		100	0.80	0.80	1.00
5	80	1	0.34	0.17	0.51
		10	0.67	0.46	0.69
		100	0.80	0.74	0.93
10	40	1	0.27	0.22	0.82
		10	0.61	0.54	0.89
		100	0.79	0.77	0.98
10	160	1	0.25	0.12	0.49
		10	0.59	0.36	0.61
		100	0.78	0.68	0.88
15	240	1	0.21	0.10	0.48
		10	0.53	0.30	0.57
		100	0.76	0.64	0.83

In a second example a balanced incomplete block design was examined in which all sires had equal progeny numbers, but that progeny of different sires were not found in the same herds. The design was connected. As may be expected large differences were observed to exist in the accuracy of each sire's proof. In warning, the calculated accuracy of a sire's proof may be greatly in error if the progeny of that sire occurred in herds in which heritabilities differ substantially from that assumed for the trait overall.

XI How to Limit Data Processing

One of the practical problems encountered when calculating BLUPs from large sets of data is that every year the problem grows larger. One method is to use only those records from the most recent time period, perhaps 10 years. Another method involves absorbing herd-year-season equations once only and storing the summed sire x sire LHS and RHS elements. This method can be used only when there are no repeated records (for example all lactations for dairy cows). It also suffers the disadvantage that in practice, error correction in previously processed data usually will reduce the expected benefits. The steady build-up of sire equations over succeeding years is not avoided.

A third and, as yet unused, method of limiting both the number of herd-year-seasons to be processed and the number of sire equations to be solved could be considered for use under the following circumstances. Suppose a dairy breeding program involves the evaluation of sires following a limited size progeny test and then extensive AI use is made of the better bulls. After about 10 years of operation, these better bulls would have very large groups of recorded progeny, while the other (unsuccessful) bulls would have no additional progeny. Retention of these unsuccessful bulls in the system of equations would be of little benefit. However, the successful bulls would probably have further progeny in many herds, and could produce connectedness between bulls in the various state testing programs and in other countries. If all data prior to an arbitrary point in time were discarded, then much of the information on these connecting bulls could be lost. On the other hand, if successful bulls were retained in the system of equations but constrained so as to have a fixed solution (that obtained with thousands of progeny) then useful prior information would be retained. It would be possible to constrain these sire solutions and in the process provide a method of "fixing" the genetic base. This method would not be desirable if a genotype x environment interaction existed and the environment was gradually changing with time. For example suppose successful bulls were not well suited to the newer environment (e.g. automatic milking) then their evaluations should be allowed to decline as more of their daughters were exposed to the new environment.

XII Conclusions

The purpose of this paper has been to draw attention to topics which we feel can either be resolved at this Symposium or be discussed at sufficient length to allow clarification of both the need for, and direction of, future research. The broad topics which we see are:

(1) How to group sires to represent genetic trend or strain differences, taking into consideration the problems of poor connectedness.

(2) How to measure strength of connectedness in a quantitative way, without matrix inversion.

(3) How to choose a genetic base - fixed or rolling.

(4) How to express BLUP results.

(5) How to allow for the c^2 effect without vast computer expense.

(6) How to choose the most appropriate BLUP method for categorical traits.

(7) How to combine BLUP estimates for an individual for various traits - economic weights versus other methods.

(8) How to allow for fixed effects - inclusion of effects in the model or preadjustment of records.

(9) How to minimize the impact of ill defined (co)variance structures on BLUP estimates.

(10) How to keep data processing costs within reasonable bounds.

(11) Finally, how can BLUP theory be extended to cater for non-linearly related traits such as milk yield and calving interval, when included in multiple trait mixed linear models.

ACKNOWLEDGMENTS

Our thanks to B.W. Wickham, H.-U. Graser, K. Hammond and C.W. Aisbett for useful discussion and comments.

REFERENCES

Berger, P.J., and Freeman, A.E. (1978). *J. Dairy Sci.* *61*, 1146

Dickinson, F.N. (1973). *Ann. Génét. Sél. Anim.* *5*, 267.

Ericksson, J.-A., Wilton, J.W., and Henningsson, T. (1979). *Acta Agric. scand.* *29*, 393.

Everett, R.W., Taylor, J.F., and Hammond, K. (1982). *Aust. J. agric. Res.* *33*, 649.

Gianola, D. (1980a). *J. Anim. Sci.* *51*, 1266.

Gianola, D. (1980b). *J. Anim. Sci.* *51*, 1272.

Hallgrimsson, S., and Jonmundsson, J.V. (1979). European Association of Animal Production 30th Annual Meeting, Harrogate, U.K., Paper GS3.8.

Hazel, L.N. (1943). *Genetics* *28*, 476.

Henderson, C.R. (1973). Proc. Animal Breeding and Genetics Symposium in Honour of Dr Jay L. Lush. Amer. Soc. Anim. Sci., Amer. Dairy Sci. Assoc. Champaign, Ill. p.10.

Henderson, C.R. (1975). *J. Anim. Sci. 41*, 760.
Miller, P. (1973). *J. Dairy Sci. 56*, 952.
Petersen, P.H. (1978). *Acta Agric. scand. 28*, 360.
Quaas, R.L., and Pollak, E.J. (1980). *J. Anim. Sci. 51*, 1277.
Quaas, R.L., and Van Vleck, L.D. (1980). *Biometrics 36*, 117.
Schaeffer, L.R. (1979). "Notes on Linear Model Theory and Henderson's Mixed Model Techniques". Mimeograph. Department of Animal and Poultry Sciences, University of Guelph, Ontario. pp. 201.
Schaeffer, L.R., and Burnside, E.B. (1974). *J. Dairy Sci. 57*, 1394.
Schaeffer, L.R., Hunt, M.S., and Burnside, E.B. (1978). *Can. J. Anim. Sci. 58*, 409.
Schaeffer, L.R., and Wilton, J.W. (1976). *J. Dairy Sci. 59*, 544.
Slanger, W.D. (1979). *J. Anim. Sci. 48*, 1070.
Slanger, W.D. (2980). *J. Anim. Sci. 51*, 595.
Taylor, J.R. (1982). Ph.D. Thesis, University of New England, Australia. pp. 231.
Van Vleck, L.D. (1966a). *J. Dairy Sci. 49*, 56.
Van Vleck, L.D. (1966b). *J. Dairy Sci. 49*, 53.
Van Vleck, L.D. (1976). *In* "Optomization of Cattle Breeding Schemes". (McGloughlin, P. ed.), p. 49. Commission of the European Communities Directorate General, Loxembourg.
Van Vleck, L.D. (1980). *J. Dairy Sci. 63*, 1172.
Van Vleck, L.D., and Hintz, R.L. (1975). Proceedings of the International Symposium on Genetics and Horse Breeding. Royal Dublin Society. Ballsbridge, Dublin, Ireland. September 17-18, 1975. p. 19.
Van Vleck, L.D. and Karner, P.J. (1980). *J. Dairy Sci. 63*, 1328.

Discussion Summary

Mixed model theory will become a more important part of animal breeding technology in the forseeable future. The progress of molecular genetics seems unlikely to affect this, although eventually it might influence the role played by mixed model theory.

Many of the methods of estimating variance components for mixed models involve starting from initial estimates of those variance components. For balanced data sets, the initial estimates have no influence, and analysis of variance, MINQUE, MIVQUE and REML all give the same estimates. With unbalanced data, the initial estimates do have an influence and the number of rounds of iteration required for convergence will vary with the data.

In selecting a variance component estimation technique for application, Anderson favoured the techniques discussed in his paper over Henderson's methods; they are often not particularly difficult to use and estimates will possess superior properties. Both covariance and variance components can be estimated. The distinction between methods which allow for the degrees of freedom associated with estimation of fixed effects and those which do not was re-emphasized. The new methods were seen as being very useful for estimating genetic correlations between traits which could only be measured on different animals (some traits on males only, others on females only).

Since rankings obtained by mixed model estimates of random effects are generally robust with respect to the variance components used, it follows that variance component estimation is not particularly crucial. However, although mixed model rankings seem robust for single traits (provided that the error variance does not vary between herds), this is not so for multiple-trait problems.

If multiple-trait rankings are to be obtained periodically, the covariance components associated with the estimated rankings should be re-estimated reasonably often, possibly every time new rankings are obtained. The population could change between estimation runs and this should be checked, even if the covariance component estimates originally used were considered to be reliable.

Grouping of bulls within mixed models should be treated with caution; to use no grouping at all was considered to be a reasonable option in many cases, providing the Numerator Relationship Matrix (NRM) is included in the model. Grouping was considered to have some application as a data snooping technique to examine connectedness between groups. The inverse of the group by group sub-matrix needs to be obtained. For models which include the NRM, grouping has been used to estimate genetic trend. Grouping was considered worthwhile for handling imported bulls (which are unrelated to the home populations but should, perhaps, be regressed towards a mean which differs from the grand mean).

The choice of genetic base and method of presentation of results were marketing decisions with few major genetic implications. There was no dissension from the protocols recommended by McClintock and Taylor.

Further research is needed on the following:

(1) Categorical traits - many problems exist in estimating genetic merit for categorical traits, which have been glossed over to date. The variance of the trait often varies with the fixed effects.

(2) Variance components - current methods for their estimation do not handle mixed models which include the Numerator Relationship Matrix, even for the single trait case. When variance components are estimated from the same data as is used for ranking animals, the properties of BLUP estimates are not obvious.

(3) Genetic correlations - their estimation is generally imprecise. Reliable genetic correlations are required for selection indices and multiple-trait BLUPs. At present, they are often the least accurate part of computations. The need for internal consistency of sets of correlations, when conducting these computations was emphasized. Perhaps a way might be found to rank animals which reduces the problems.

(4) Within-sire residuals - it has been suggested that some sires "stamp their type" on their offspring more clearly than other sires. This theory could be considered by looking at residuals from a mixed model, or some similar technique. If within-sire residuals do differ, then new models might be required.

Part 7
Population Size

Population Size and the Genetic Improvement of Animals

I. R. Franklin

CSIRO Division of Animal Production
Genetics Research Laboratories
North Ryde, New South Wales

In commercial breeding programs, both effective number and the size of the recorded population are determined by a variety of genetic and economic factors. There is a minimum size, primarily dictated by inbreeding considerations, but optimum size is usually defined by a compromise between anticipated costs and the theoretical genetic gains. This paper argues that there are both genetic and managerial reasons to suppose that predicted gains from intense selection are rarely achieved in practice in large programs, and advocates smaller, well designed, improvement schemes.

I Introduction

By almost any criterion, population size is a crucial variable in the design of animal breeding programs. First of all the size of the recorded population places constraints on the selection intensity, and hence the expected response, for each of the traits under selection. Second, as we are all aware, small populations lead to inbreeding, with all of its attendant problems of decline in vigour and reproductive capacity. Added to this we have variability in response, fixation of 'deleterious' alleles, and theoretically, a smaller cumulated response at the selection limit. Against this must be balanced the very important factor of cost, for large programs may be very expensive to establish and

FUTURE DEVELOPMENTS IN THE
GENETIC IMPROVEMENT OF ANIMALS
ISBN 0 12 078830 6

administer. Therefore, for the reasons alluded to above, many quantitative geneticists advocate that breeding programs should be as large as possible (e.g., Harris, 1977). An extreme example is the intensive dairy industry, where the costs of recording can be justified independently of genetic improvement. Here selection programs involving in excess of one million cows and very large selection differentials can be found.

I believe that this common attitude that larger breeding programs are better than small programs is based on dubious assumptions, and that cost-benefit analyses involving intense selection may be in error. There is no doubt that there is good general agreement between the theory and observation for short term response in experimental populations, especially for single trait selection. However, the application of quantitative genetics to animal breeding requires the extrapolation of these experimental results to large scale, and often intense, multitrait selection. Added to this we are faced with considerable statistical and managerial problems which do not arise in experimental populations. The major purpose of this paper is to question whether selection response can be expected to follow theoretical prediction as the size of breeding programs increases.

II The Theoretical Basis of Selection Response

The theory of selection response for continuously variable traits has been built around assumptions of infinite population size and constancy of heritabilities and genetic correlations. In the short term there is experimental and theoretical (Kimura, 1958) support for this, although genetic correlations are thought to be more labile than heritabilities. Underlying this theory, of course, is the assumption that quantitative traits are controlled by a large number of loci, and that alleles at each locus have a small effect. Biologically this makes little sense, and if we are to have a truly predictive theory of selection response it is necessary that we have a greater understanding of the developmental processes which underly quantitative variation.

It may be that in a relatively unselected population, the properties which we usually associate with quantitative variability are a reflection of a long history of natural stabilising selection, under which only alleles with small effect can reach an appreciable frequency (Fisher, 1930).

If this is the case the 'genetic architecture' of quantitative traits is not a fundamental property of that trait but simply a function of selection history. Under directional selection the genetics of the trait may change considerably, particularly if selection is intense. Alleles with large positive effects, even recessive lethals, which were initially rare or which arise by mutation will now be favoured, and such alleles are often noted in long term selection experiments.

Ignoring for the moment these reservations about the nature of quantitative variability, I shall now turn to the theory of selection in small populations.

A. *Selection Response in Finite Populations*

It was not until 1960 that some of the consequences of finite population size were analysed in Robertson's important paper on limits to selection. Subsequently there have been theoretical contributions by Hill and Robertson (1966), Robertson (1970), Avery and Hill (1977) and Nicholas (1980) discussing various aspects of sampling effects on selection response, and numerous experimental studies, notably those of Jones *et al.* (1968), Hammond (1973), Frankham (1977) and Yoo (1980) with *Drosophila* and Roberts (1966), Falconer (1973) and Eisen (1975) in the mouse.

In order to set a framework for subsequent discussion, I shall briefly review the theory of selection limits and optimisation of selection response.

Suppose that we have a selection program in which each generation there is a constant selection differential i, and that an effective size N is maintained throughout. Let the phenotypic and additive genetic variances in the kth generation be $V_P^{(k)}$ and $V_A^{(k)}$ respectively, then the response in the k+1th generation is

$$\Delta G^{(k)} = \frac{i\, V_A^{(k)}}{\sqrt{V_P^{(k)}}} \quad .$$

If there are no dominance or epistatic contributions to the genotypic variance, the additive variance will be reduced by a proportion (1 - 1/2N) each generation, i.e.

$$V_A^{(k)} = V_A^{(0)} \left(1 - \frac{1}{2N}\right)^k \quad .$$

Hence,

$$\Delta G^{(k)} = \frac{i}{\sqrt{V_P}^{(k)}} V_A^{(0)} \left(1 - \frac{1}{2N}\right)^k \qquad (1)$$

$$\simeq \Delta G^{(0)} \left(1 - \frac{1}{2N}\right)^k,$$

assuming that V_P remains constant. The total response to the tth generation is

$$\sum_{k=0}^{t-1} \Delta G^{(k)} \simeq \frac{i\, V_A^{(0)}}{\sqrt{V_P}} \sum_{k=0}^{t-1} \left(1 - \frac{1}{2N}\right)^k$$

$$= 2N\Delta G^{(0)} \left[1 - \left(1 - \frac{1}{2N}\right)^t\right]. \qquad (2)$$

Therefore as t increases, the selection limit is approximately 2N times the response in the initial generation.

It can be seen that the selection limit is a function of Ni. Suppose that T is the size of the recorded population, and that a proportion p is selected each generation. If N depends only on the number of parents, Tp, then

$$Ni \propto Tpi = Tp\, z/p = Tz,$$

where z is the ordinate of the distribution at the truncation point and is a maximum at the mode of the phenotypic distribution. If this distribution is symmetric (e.g., Normal) the proportion selected which maximises eventual gain is 0.5.

Another criterion put forward as important when designing selection programs is the ultimate fixation probability for a favourable gene with initial frequency q (Comstock, 1977). This is, for additive genes,

$$u(q) = (1 - e^{-2Nsq})/(1 - e^{-2Ns}).$$

For genes of very small effect α, the selection differential s is a constant, proportional to $i\alpha$, and, therefore, it is possible to calculate the probability of fixing an allele with an effect, say, of 1%, and initial frequency 0.1, for a range of effective population sizes. Using an argument similar to the above it is possible to demonstrate that for a given initial gene frequency, u(q) is maximised when the proportion selected is one half.

This then is selection limit theory in its simplest form. The theory predicts that in the long term the greatest cumulated response is achieved under weak selection ($p = 0.5$), but that in the short term, response is independent of population size and proportional to the selection differential.

Equations (1) and (2) were used by Robertson (1970) to derive optimum selection intensities when a compromise needs to be made between short term and long term genetic gains. These equations were also modified by James (1972) to include discount rates and hence derive the present value of future gains. From these equations he derived optimum values for selection intensities in terms of initial and current costs, future returns, and population size.

All of these optimisation questions depend on the assumption that additive variance declines at a rate $(1 - 1/2N)$ per generation. In the absence of knowledge of gene effects, initial frequencies and linkage relationships we cannot predict the effect of selection on additive variance. Perhaps more important is the fact that Equations (1) and (2) do not allow for renewal of genetic variance by factors such as mutation and recombination. Frankham (1980) has argued that each of these factors are important in long term selection response.

B. *Selection Response with Renewal of Additive Variance*

Suppose that the renewal rate is constant and equal to V_m per generation. Given an initial variance V_A, we have after one cycle of selection

$$V_A^{(1)} = V_A^{(0)}(1 - \frac{1}{2N}) + V_m.$$

After two generations

$$V_A^{(2)} = (V_A^{(0)}(1 - \frac{1}{2N}) + V_m)(1 - \frac{1}{2N}) + V_m$$

etc.

Hence

$$\begin{aligned} V_A^{(k)} &= V_A^{(0)}(1 - \frac{1}{2N})^k + \sum_{j=0}^{k-1} V_m(1 - \frac{1}{2N})^j \\ &= V_A^{(0)}(1 - \frac{1}{2N})^k + 2NV_m[1 - (1 - \frac{1}{2N})^k] \end{aligned}$$

therefore

$$\Delta G^{(k)} \simeq \frac{i}{\sigma} V_A^{(0)} (1 - \frac{1}{2N})^k + \frac{2Ni}{\sigma} V_m [1 - (1 - \frac{1}{2N})^k] \quad (3)$$

where $\sigma = \sqrt{V_P}$.

The response is separable into two components, one due to the initial additive variance, and the other a function of V_m. The population responds indefinitely, with the second term becoming increasingly important. At the limit

$$\Delta G^{(\infty)} = 2Ni\, V_m/\sigma \quad (4)$$

and the asymptotic additive variance is

$$V_A^{(\infty)} = 2NV_m.$$

The cumulated response to the tth generation is

$$\sum_{k=0}^{t-1} \Delta G^{(k)} \simeq \sum_{k=0}^{t-1} \frac{iV_A^{(0)}}{\sigma} (1 - \frac{1}{2N})^k + \sum_{k=0}^{t-1} 2Ni \frac{V_m}{\sigma} (1 - (1 - \frac{1}{2N})^k)$$

$$= \frac{2NiV_A}{\sigma} [1 - (1 - \frac{1}{2N})^t]$$

$$+ \frac{2NiV_m}{\sigma} [t - 2N(1 - (1 - \frac{1}{2N})^t] . \quad (5)$$

If $t >> N$, the total response becomes

$$\frac{2NiV_A}{\sigma} + \frac{2NiV_m}{\sigma} (t - 2N).$$

There are a number of possible causes for renewal of variability during the course of selection, such as mutation, recombination or migration. In the analysis above it was assumed that there is a constant increment to the additive variance by an amount V_m per generation, with no average effect on the mean. This may be appropriate for mutation or unequal crossing over (Frankham *et al.*, 1978). In the case of mutation, we expect each generation an increase in variance

$$V_m = 2 \sum_{\text{all loci}} \mu_i m_i^2$$

where μ_i is the mutation rate at the ith locus, and m_i is the average effect of new alleles at this locus. For *Drosophila* bristles this quantity appears to be of the order of 10^{-3} times the environmental variance per generation (see Lande, 1976).

In order to illustrate the effect of generation of variability by mutation or unequal crossing over, Figure 1 shows the cumulated response (Equation 5) for three effective population sizes and a range of V_m values. While mutation has little noticeable effect up to about 20 generations, there is an increasing contribution from mutational variance thereafter. The limiting response is directly proportional to N, and hence is greater for large populations, but the effect of mutation is relatively greater, and hence more noticeable, for small populations. Since it appears that $V_m = 10^{-3}$ is appropriate for bristle variation in *Drosophila*, the contribution of mutation may explain the continued response to selection at 50 generations in the experimental results of Jones *et al.* (1968).

Hill (1982) used fixation probabilities of newly arising genes to derive the contribution of mutation to selection response, and as expected, the conclusions for additive genes do not differ from those derived above. The formulation used here, and its justification, is one that I have used before (Franklin, 1980) in a consideration of minimum effective size for genetic conservation.

It is clear that there are some theoretical objections to this simple model. The equilibrium additive variance is $2NV_m$, and therefore if $N > 1/(2V_m)$, the additive variance, and hence heritability, will increase over the course of selection. While some selection experiments do not exclude this possibility, there are major evolutionary difficulties in assuming that observed heritabilities are a reflection of an equilibrium between mutational variance and effective size, as would be the case for neutral variation. (If V_m is $10^{-3}V_e$ for bristle genes, and $h^2 = 0.4$, the implication is that the effective size of *Drosophila* populations has been about 300 over recent evolutionary history.) Although there are clearly factors other than mutation and effective size contributing to the maintenance of variability for quantitative traits it is hoped that the above approach provides an indication of the role of mutation in selection response.

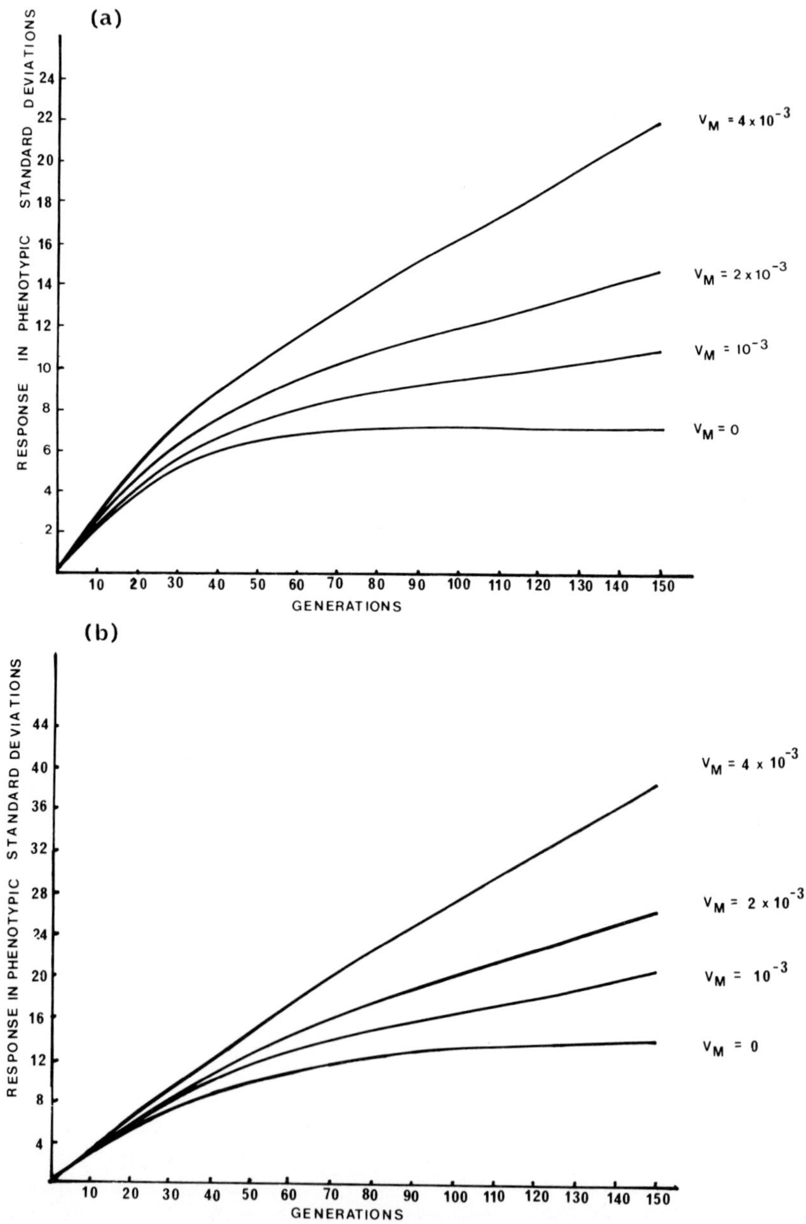

FIGURE 1. The cumulated response to selection for an initial heritability of 0.25, selection differential $i = 1.4$ ($p = 20\%$) and $V_m = 0$, 10^{-3}, 2×10^{-3} and 4×10^{-3}. (a) $N = 10$. (b) $N = 20$. (c) $N = 40$.

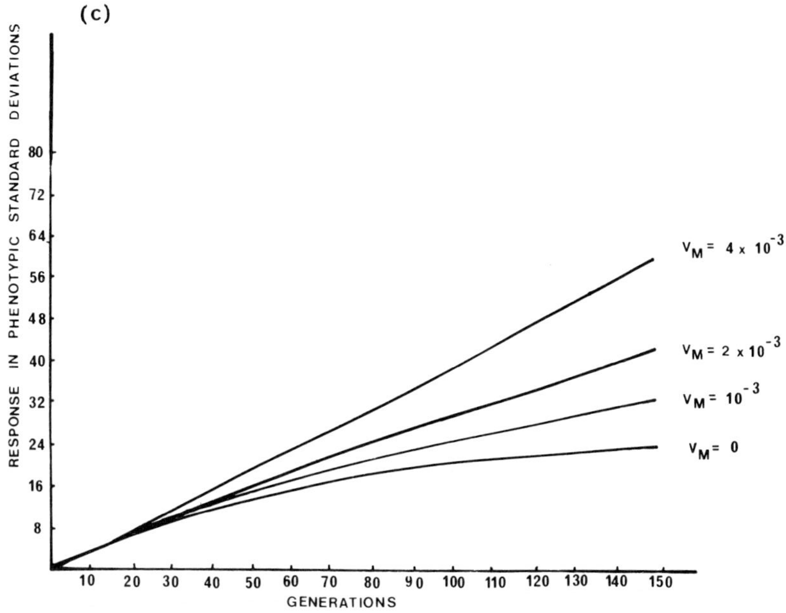

FIGURE 1,c.

Another source of variance is the break up of repulsion linkages which may have built up over long periods of stabilising selection. Such a source of variability would be a function of the degree of heterozygosity and recombination parameters, and may be expected to decline over the course of long term selection as the gametic frequencies find new equilibrium values.

The significance of this theory is that if V_m is sufficiently large there is effectively no limit to response, but that the cumulated gain in the long term is still a function of Ni. For a given total recorded size (T), this quantity is still maximised when $p = 0.5$.

III Population Size and Animal Improvement

Having cursorily examined the theoretical background to selection response in finite populations, I would like to turn to the question of optimum population size in practical breeding programs. Here the situation is much more complex, and because of the long generation intervals of commercial

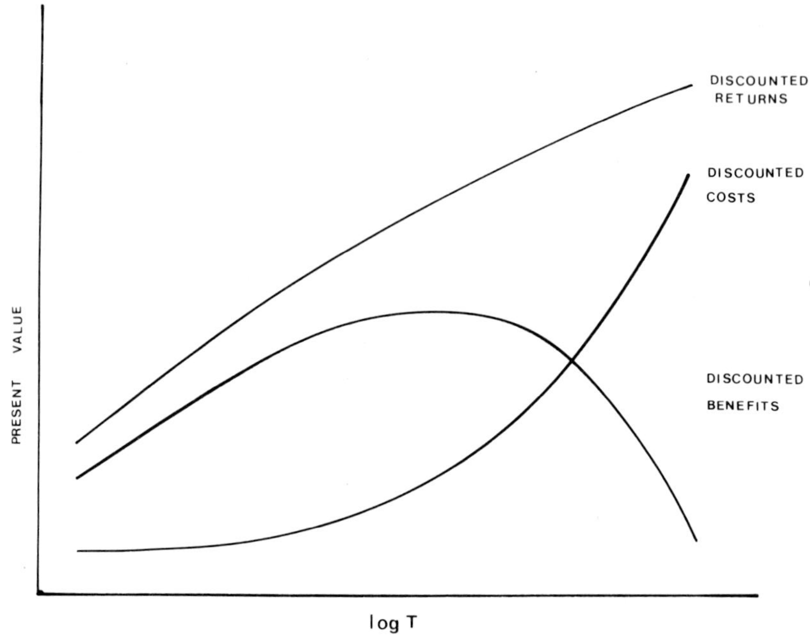

FIGURE 2. Costs, returns and benefits as a function of population size. From Smith (1981).

species, selection limit theory has little relevance to the industry. In fact, since animal breeding programs necessarily involve multitrait selection, it is probably impossible to derive useful expressions for long term gains. While the constancy of heritabilities has theoretical and empirical support, the assumption of constancy for genetic correlations has far less justification.

First and foremost in the design of breeding programs is the question of benefit to the industry, and one of the most popular ways of assessing this is to compare predicted returns against predicted costs, each appropriately discounted. Following James (1972), Smith (1981) suggests that the relationship between costs, returns and benefits has the form shown in Figure 2, namely that costs are directly proportional to the size of the recorded population (T), whereas returns are roughly linear in log(T). As a result each breeding program has a maximum benefit at some intermediate value of total population size. In the intensive dairy industry, recording costs are largely irrelevant, and in this case theory tells us that programs should be as large as possible, with very high selection intensities. Such a conclusion is not necessarily true for Australia, where recording has much less value for management decisions.

This cost-benefit approach makes a fundamental assumption that returns rise steadily as programs become larger. Before focussing on questions of optimum size it is clear that breeding programs have a minimum size.

A. *Minimum Population Size*

The minimum effective size of a breeding program is dominated by inbreeding considerations. This does not necessarily mean that a new program should be initiated at an effective size of say 50, but a value such as this should be attained within a few generations. If the program proves successful this should not be difficult.

Another criterion for minimum size is predictability of response. The effective size should be large enough that some reasonable fraction of the expected response is attained with a probability of say 90%. This kind of criterion is important if only to justify the cost-benefit analysis. Using the approximate theory of Avery and Hill (1977), Nicholas (1980) considered these questions in detail, and I shall discuss them no further.

In addition to these genetic considerations there are economic factors in the minimum size of a breeding program. Programs should be large enough to generate enough income through sale of livestock, embryos or semen to meet their costs.

B. *The Relationship between Response and Selection Intensity*

It is assumed that by keeping N constant, but increasing the total recorded population, the response per generation increases as a direct function of the theoretical selection intensity. I believe that there are both genetical and managerial reasons to be suspicious of such a proposition. The genetical assumption of course is that the regression of offspring on parent remains linear throughout the entire range of parental values. When working at the extremes of distributions we may find that predictions made from an analysis of the base population may no longer hold true. An important possibility is that genotypic and environmental effects may not be identically distributed,

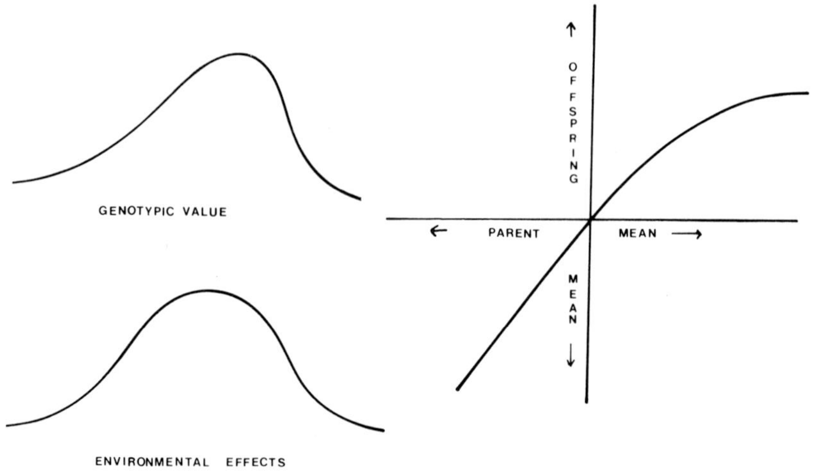

FIGURE 3. The effect of differences in scale for genetic and environmental effects on linearity of offspring-parent regression.

and in this case non-linearity is expected, as pointed out in an important paper by Nishida and Abe (1974). This phenomenon is illustrated in Figure 3, where offspring means do not increase linearly as parents are chosen from a smaller and smaller proportion of the top end of the distribution. An example of different distributions for genetic and environmental effects exists for pupa weight in Tribolium, where I found (Franklin, 1967) that on the genetic scale a log transformation was appropriate for equalisation of variances, but a cube root transformation was required to equalise environmental variance. Of course such non-linearity may work to the advantage of the selection program, and this could be checked by measuring response to up and down selection. Regardless of the cause, it appears that asymmetry of response is not uncommon (Robertson, 1977); a fact which is not easily explained by the classical models of quantitative inheritance.

Another potential factor in non-linearity of response is the presence of certain alleles which play an important role in the development of the trait. For example, the presence or absence of the 'scute' gene in *Drosophila* selection lines makes an enormous difference to the predicted response in either the upward or downward direction, for in this case the genetic variability which is available in the presence of sc^+ seems to have little effect in the presence of scute mutations, and vice versa.

Our fundamental ignorance of the underlying genetic basis for quantitative traits should in general make us wary of selecting among extremes as I suggested earlier. We may for example be choosing individuals in which there has been a basic deregulation of an important developmental process, and these may be associated with undesirable physiological effects. Often in long term selection lines one finds homozygous lethals which have been selected for their heterozygous effect on the trait, and their frequency may be high if selection is intense. While the search for exceptional phenotypes may be to our advantage, we cannot always expect genetic deviants to follow the rules of more conventional quantitative inheritance.

C. *Managerial Limitations in Large Breeding Programs*

I believe that the most important factors in the failure to achieve theoretical gains, especially in large programs, are problems of management and analysis. A simple example is the presence of heterogeneities of variance in herds or studs, for if these go undetected the algorithms which detect superior animals will preferentially select animals from the more variable herds.

As we increase the size of the recorded population we necessarily spread the recorded population over a larger sample of environments, and if we are to take advantage of this larger population the statistical models which are designed to remove fixed effects such as season and age effects become more complex. Hence it becomes difficult to remove extraneous variation as the environments become more diverse, and deficiencies in the models become more critical if we are attempting to identify extreme individuals, rather than identify say the top 10% of a herd or flock.

Another major problem is the occurrence of errors in pedigree, especially amongst 'elite' animals. In effect such errors reduce the heritability. This would not militate against large programs if such errors were independent of the size of the population, but I do not believe that this is the case. It is, I maintain, much easier to administer and monitor small programs with a small staff each of whom are aware of the aims and the importance of accurate records.

There are perhaps only two industries, namely poultry and dairy, where good estimates of genetic progress in non-experimental populations are available. Estimates of genetic progress in milk yield in North America (Van Vleck, 1977) are of the order of 0.5% per annum, which is far short of the theoretical maximum. In contrast, the AMZ tropical breeding program under the control of CSIRO has a cow population of less than one thousand, yet gains of 1.7% per annum are being achieved for the sire population (Everett, Franklin and Hammond, 1982). This rate is close to the theoretical prediction. In poultry, Nordskog (1977) expressed general dissatisfaction with rates of genetic improvement in egg production, although Kashyap, Dickerson and Bennett (1981) reported improvement in carefully managed commercial flocks which agreed with predicted rates. Both industries illustrate that theoretical gains can be achieved under rigid statistical and managerial control, but often fall far short in practice.

IV Conclusions

When designing animal breeding programs quantitative genetic theory tells us that the size of the recorded population and the selection intensities should be as large as possible, and constrained only by the costs involved in recording and management. The point of the above discussion is to argue that 'optimum' population sizes may be less than we suppose. It is possible that in poorly designed programs response may actually decrease as the size of the recorded population increases, through failures in discipline and increased complexity of the analysis.

I have also suggested that the importance of population size on the anticipated limits to selection may not be as great as previously indicated, and that there may be genetical reasons to be wary of high selection intensities. In the past selection intensities have been limited by the reproductive capacity of the large animal species, but this constraint is now being removed by the increasing use of AI and ovum transplantation.

I believe that there is in Australia an obsession with size, as indicated by the pressure for national breeding programs instead of small, well designed local ones, and that such programs may be counter productive in the extensive Australian animal production industry. It may be that large programs are appropriate in the more intensive large animal industries of Europe and North America, where environmental variables may be well controlled, but this is not necessarily the case in Australia. In developing countries, where there is a shortage of trained technicians, the advocacy of large breeding programs is I feel irresponsible.

There are a few final points based on evolutionary theory and concepts such as adaptation. In the presence of non-additive variation, Wright (1931) argued that adaptive peaks exist which will limit progress unless the populations are split into a large number of semi-isolated groups. If this concept is valid for evolutionary progress I see no reasons why similar constraints should not apply in animal breeding. Another factor is the importance of genotype by environment variance. Local adaptation is well recognised in plants - no one would advocate the establishment of a national breeding scheme dedicated to the improvement of a single Australian cereal variety - but is less accepted in animal breeding. These unresolved questions of non-additive variation are further arguments for fragmentation rather than amalgamation.

ACKNOWLEDGMENTS

I am grateful for Drs. B. McGuirk and R. Frankham for helpful suggestions and criticism.

REFERENCES

Avery, P.J., and Hill, W.G. (1977). *Genet. Res. 29*, 193.
Comstock, R. (1977). *In* "Proceedings of the International Conference on Quantitative Genetics" (E. Pollak, O. Kempthorne and T.B. Bailey, eds.), p. 705. Iowa State Univ. Press, Ames.
Eisen, E.J. (1975). *Genetics 79*, 305.
Everett, R.W., Franklin, I.R., and Hammond, K. (1982). *In preparation.*
Falconer, D.S. (1973). *Genet. Res. 22*, 291.
Fisher, R.A. (1930). "The Genetical Theory of Natural Selection." Clarendon Press, Oxford.
Frankham, R. (1977). *Genet. Res. 30*, 115.

Frankham, R. (1980). *In* "Selection Experiments in Laboratory and Domestic Animals" (A. Robertson, ed.), p.55. Commonwealth Agricultural Bureau, Farnham Royal.
Frankham, R., Briscoe, D.A., and Nurthen, R.K. (1978). *Nature, Lond. 272,* 80.
Franklin, I.R. (1967). Ph.D. Thesis, University of California, Berkeley.
Franklin, I.R. (1980). *In* "Conservation Biology" (M.E. Soulé and B.A. Wilcox, eds.), p. 135. Sinauer Associates, Massachusetts.
Hammond, K. (1973). Ph.D. Thesis, University of Sydney.
Harris, D.L. (1977). *In* "Proceedings of the International Conference on Quantitative Genetics" (E. Pollak, O. Kempthorne and T.B. Bailey, eds.), p. 587. Iowa State Univ. Press, Ames.
Hill, W.G. (1982). *Proc. Natn. Acad. Sci. U.S.A.,* (in press).
Hill, W.G., and Robertson, A. (1966). *Genet. Res. 8,* 269.
James, J.W. (1972). *Anim. Prod. 14,* 1.
Jones, L.P., Frankham, R., and Barker, J.S.F. (1968). *Genet. Res. 12,* 249.
Kashyap, T.S., Dickerson, G.E., and Bennett, G.L. (1981). *Poult. Sci. 60,* 1.
Kimura, M. (1958). *Heredity 12,* 145.
Lande, R. (1976). *Genet. Res. 26,* 221.
Nicholas, F.W. (1980). *Genet. Res. 35,* 85.
Nishida, A., and Abe, T. (1974). *Can. J. Genet. Cytol. 16,* 3.
Nordskog, A.W. (1977). *In* "Proceedings of the International Conference on Quantitative Genetics" (E. Pollak, O. Kempthorne and T.B. Bailey, eds.), p. 569. Iowa State Univ. Press, Ames.
Roberts, R.C. (1966). *Genet. Res. 9,* 73.
Robertson, A. (1960). *Proc. R. Soc., Ser. B153,* 234.
Robertson, A. (1970). *Theor. Populat. Biol. 1,* 120.
Robertson, A. (1977). *In* "Proceedings of the International Conference on Quantitative Genetics" (E. Pollak, O. Kempthorne and T.B. Bailey, eds.), p. 297. Iowa State Univ. Press, Ames.
Smith, C. (1981). *Livest. Prod. Sci. 8,* 193.
Van Vleck, L.D. (1977). *In* "Proceedings of the International Conference on Quantitative Genetics" (E. Pollak, O. Kempthorne and T.B. Bailey, eds.), p. 543. Iowa State Univ. Press, Ames.
Wright, S. (1931). *Genetics 16,* 97.
Yoo, B.H. (1980). *Genet. Res. 35,* 1.

The Significance for Genetic Improvement of the Number of Individuals Available for Breeding

Keith Hammond

Animal Genetics and Breeding Unit[1]
University of New England
Armidale, N.S.W.

Current developments in reproductive biology and in estimating genetic worth could be combined to increase substantially the number of animals available for selection and reduce the number selected as parents. The very high selection intensities which current theory suggests as being desirable to maximize short-term gains could be achieved. A more detailed examination of the possibilities is proposed to clarify the value of such changes in the design of breeding programs.

I Introduction

The number of individuals available for selection and the number selected are important variables governing the rate of, and total response to, recurrent selection in animal breeding and the variation about predicted response. They have been considered in recent discussions of selection theory by Bohren (1975), Harris (1977) and Eisen (1980). Here, I simply wish to highlight an aspect which may be of use to achieve increased responses in current breeding programs but which first requires further study. The aspect

[1] *A joint Unit of the New South Wales Department of Agriculture and the University of New England.*

FUTURE DEVELOPMENTS IN THE
GENETIC IMPROVEMENT OF ANIMALS
ISBN 0 12 078830 6

relates the current developments in reproductive biology and in estimating an individual's genetic worth, both referred to earlier at this Symposium, to the number of potential and of selected parents prior to each mating season.

II Prediction of Responses

Until recently, the classical additive model of Dickerson and Hazel (1944) and Rendel and Robertson (1950) has been used to predict the response (R) in the next generation and to maximize or optimize the rate of response in the short-term, in the for:

$$R = r_{IA} \, i\sigma_A$$

where r_{IA} is the correlation between the true additive genetic value and the estimated additive genetic value for the selection criterion, frequently termed the accuracy of selection.

i is the selection intensity (in standard units), and

σ_A is the additive genetic standard deviation of the selection criterion.

After summing R over the two (individual selection) to four (progeny testing programs) pathways for transmission of genetic improvement and doing the same for the generation intervals, the annual change in the selection criterion has been obtained. Strictly, the model applies to an infinite population but various provisions can be made for finite numbers, viz.

(1) *Genetic variance* - where a small founder sample (n) has been taken at random to initiate the breeding program from a large population on which σ_A was estimated, James (1971) predicted that R would be reduced by $1/2n$ and Hammond (1974) provided experimental evidence confirming this. In addition to selection depleting σ_A (for which there is inadequate theory), drift will reduce σ_A^2 each generation by $1/2N$, where N is the effective population size (e.g., James, 1972, and see Enfield, 1977 for experimental evidence). Of course drift will also influence the standard error of prediction for R (see Hill, 1977a for a recent discussion of the adequacy of the theory); this influence is frequently underrated when responses to selection are being interpreted.

(2) *Selection intensity* - with finite numbers of potential parents and of parents selected each generation, i is reduced (Falconer, 1960). Appropriate values can be derived from normal distribution theory and these have been tabulated (Becker, 1974).

(3) *Inbreeding depression* - the selection criterion may include both fitness and production traits. Where genes affecting these show directional dominance or epistasis, R will be reduced by $D/2N$ per generation where D is the loss in performance in a completely inbred population and the rate of inbreeding (assuming random mating) is $1/2N$.

As an indication of the cumulative effect of these factors on mean response, consider the following example. The owners of two herds of 60 and 150 cows decided to initiate a breeding program by each purchasing one set of founder bulls for use in the first two years, after which the herds were closed with each replacement bull being used only once at 18 months of age on 30 females. The best 10% of bulls were selected as replacements and the selection criterion suffers depression of 1% per 1% rate of inbreeding. Table I

TABLE I. Two Examples of Approximate Depression (%) in Rate of Response to Selection Resulting from Finite Numbers Available for Breeding[a]

	Herd size	
	2 bulls+60 females	*5 bulls +150 females*
Founder size	6.25	2.25
Selection intensity	6.7	2.8
Drift	2.5	1.0
Inbreeding depression	2.5	1.0
Total depression in second generation	17.95	7.05

[a]*The effective founder population size (n) of the small and large herds is 8 and 20, respectively; the corresponding effective sizes (N) in subsequent generations are 20 and 50, respectively.*

summarizes the approximate depression in response expected in the two herds relative to the response obtained in a very large herd with otherwise the same structure. Assuming that effective population size is given by four times the number of males used per generation and that the above asymptotic results hold early in the program, we expect a loss in response in the second generation and thereafter of 17.95% and 7.05% in the 60-cow and 150-cow herd, respectively.

With separate sexes, N is often taken as four times the number of males used per generation (see James, 1978 for a detailed discussion of formulae for N). However, with selection (Robertson, 1961) and a non-Poisson distribution of life-time family size when generations overlap (Hill, 1977b), N can be appreciably overestimated using this approximation.

Although the classical model has been used for both discrete and overlapping generations, in the latter structures candidates for selection are progeny of parents of different ages differing in genetic merit, and hence belonging to separate distributions of the selection criterion (Bichard *et al.*, 1973). Further, in the early years of the selection program, gains will accumulate slowly. To predict the response prior to and at the steady-state, account should be taken of the age structure of the population (see Hill, 1977b, for a review of approaches taken) and of these separate distributions (Hopkins and James, 1979a). Response will be maximized when all available animals are used as potential parents for the next mating. Hopkins and James (1977) termed this 'parent selection'. They developed theory to predict the contribution each age group should make to the breeding population when the genetic differences between age groups are at equilibrium (Hopkins and James, 1979a). When comparing the impact of this parent selection with progeny selection, Hopkins and James (1977) noted that the former led to an increase in σ_A available at selection. Hopkins (1978) commented that the resultant increase in selection efficiency may result in N being reduced, with considerations of inbreeding depression and reduced genetic variance in future generations (Hopkins and James, 1979b) becoming more important in determining optimum designs.

An approximate theory also exists to predict the maximum (but see Hill, 1982, and Franklin, 1982 for the incorporation of mutation) total response and approach to the limit (Robertson, 1960). Again N is omnipresent, the limit being $2N$ times the response in the first generation and the half-life $1.4N$ generations. However, laboratory experiments (Jones *et al.*, 1968; Falconer, 1977; Eisen, 1980; Yoo, 1980) and examinations of opposing natural selection (James, 1962; Nicholas and Robertson, 1980), suggest that we should expect half-lives to be actually about $0.5N$ generations.

This limits theory and the experimental work stimulated by it, have helped to provide an appreciation of the remarkable plasticity of quantitative genetic variation and insight into its nature (for a discussion, see Falconer, 1977) and into the means of overcoming selection plateaux. Nevertheless, contrary to the views of Comstock (1977), Harris (1977) and Eisen (1980), the limits theory is of little direct concern in the planning of recurrent selection programs in livestock, because even $0.5N$ generations will generally be too long a period. Breeding objectives and hence selection criteria will change with changes in market forces and production systems. Breeding interests will generally be concerned with the short-term benefits. Creating the "ideal genotype" by fixing the most useful allele at every locus contributing to the variation in the selection criterion (Comstock, 1977) is of minor interest. Consequently, the approach of Smith (1969) - maximum biological gains in a specified time horizon (t) - and of James (1972) - maximum present economic value in the time horizon (t) - is preferred.

The results of these studies of optimum procedures for steady-state mass selection by Smith and James, together with further work by Smith (1981), suggest that for small values of t (say t = 5 generations), at acceptable interest rates (e.g., costs 3% and benefits 5% - Bird and Mitchell, 1980) and with large numbers available for selection, the optimum selection intensity should be very high compared to values currently employed. In these circumstances the rate of response will be limited only by the level of inbreeding and the reproductive rate (Smith, 1981). Smith (1969) accounted for both loss of genetic variance due to drift and of performance due to inbreeding (but excluding inbreeding effects of selection), whilst James (1972) included only the former and Smith (1981) neither.

III Recent Developments

With the increasing application of developments in reproductive biology, particularly in cattle, sheep and pigs, the constraints on i imposed by natural means of reproduction are being reduced.

Simultaneously the development and application of BLUP methods (for a description see Henderson, 1977) is rapidly expanding (for example, see McClintock and Taylor, 1982). In some populations it is now possible to rank on Estimated Breeding Value (EBV) all recorded animals (parents and progeny) across all generations. The major advantage of BLUP for the genetic improvement of livestock is not in its ability to increase r_{IA} over previously used methods (by being that linear predictor of breeding value which has minimum variance and is unbiased), but in what it offers to the processing of performance records and to other aspects of the design of breeding programs, viz.

(1) Ability to produce EBVs for all past, present and potential sires and dams on file from the one analysis.

(2) Potential to account for selection on performance at earlier ages.

(3) Flexibility in and ease of producing multiple-trait indexes.

(4) The capability of producing annual herd/flock reports detailing genetic and environmental trends.

As the BLUP methodology permits estimation of breeding values for all potential parents for the next mating, removing genetic differences between age groups resulting from selection, it has important ramifications for the design of breeding programs. It means that the use of the Hopkins and James (1977) prediction of the proportion to be selected from each parent age group can be replaced wherever it is possible to apply BLUP methods (James, 1979). The potential parents for the next mating are all animals on hand, both those present in the paddock or represented as frozen semen or embryos. The resulting increase in numbers available for selection removes the need to emphasize generation interval when implementing breeding programs, as age is no longer used in estimating genetic trends.

The combination of high reproductive rates and BLUP methods will provide the potential for intense selection and small effective population size. Future possibilities of storing semen of sheep and pigs, of sex control and of cloning will permit even higher reproductive rates. The large scale progeny testing programs currently used for cattle may become feasible in other species.

IV What is Required?

Approaches such as those of Hill (1974) and Hopkins and James (1979a,b) need to be extended and applied to specific structures for a range of relevant variables and for individual and family selection regimes. As Smith (1981) points out, after recognizing a similar need following his recent study of optimum levels of investment, this is not a trivial exercise.

It is likely that the optimum designs suggested by these studies will differ significantly from the currently accepted norms. In the large scale progeny testing programs, the relative importance of the four pathways of gene transmission is likely to differ from, for example, those determined by Skjervold (1963) for a range of situations, particularly for the smaller populations (containing, say, less than 50,000 performance recorded and artificially bred cows). The suggestion by Dempfle (1975) to select sires of sons within half-sib families, to increase N without severely reducing early gains, may well prove advantageous.

Such studies need to account for inbreeding depression, including the inbreeding effect of selection, and possibly for reduction in genetic variance. Several other variables, seldom included in past studies, should eventually be added because of their practical significance. For example, the majority of the past empirical studies of design have been deterministic and have not incorporated random events in the field such as variation in sex ratio, survival and herd size. These can have substantial impact on the actual design (Hinks, 1974; Hudson *et al.*, 1979). For example, in Table II are the results of a small simulation in which sex ratio and survival were varied for 20, 30, 40 and 80 bulls under progeny test and where the aim was to obtain 30, 30, 40 and 100 first lactation records per bull, respectively. Semen may not be purchased from those bulls which by chance have obtained considerably fewer progeny records because of the higher risk involved. Thus, when implementing a progeny testing program emphasis is given to the minimum number of progeny per bull rather than to a mean. At the same testing capacity fewer bulls can then be evaluated. Alternatively, the progeny testing authority may decide to use the mean number of progeny when determining the number of inseminations required per bull and discard those bulls with insufficient recorded progeny. The capacity to progeny test for sex-

TABLE II. *A Simulated Distribution of Numbers of Bulls According to Progeny Group Size for Varying Numbers of Bulls Tested, of Inseminations per Bull and Progeny Survival*

Number of bulls tested	Progeny survival[a] (birth till completed first lactation)	Inseminations per bull[a]	0–9	10–19	20–29	30–39	40–49	50–59	60–69	70–79	80–89	90–99	> 100	Aim	Min.	Mean	Max.
20	0.6	200		1	9	9	1							30	18	29	40
30	0.6	200		1	17	9	3							30	19	30	42
40	0.5	300			4	15	19	2						40	27	39	53
80	0.5	800								2	13	25	40	100	74	99	122

[a]Progeny survival, inseminations per bull and sex ratio assumed to be binomially distributed with average inseminations per live calf being 2.0 and average sex ratio of 0.495 females.

limited traits will be substantially increased once sex control is feasible, particularly in small populations where variation in sex ratio is more of a problem.

Further, the sizes of livestock industries change over time and the present value of a breeding program will depend on the size of the future population over which its costs and returns are spread. For example, Brascamp (1973) found that the value of genetic improvement in a population decling in size at a constant rate d per year over 25 years was $(1-d)^{16}$ times the value with population size constant. Such changes in size are often unpredictable but a range of values should be examined.

Finally, these developments also have important ramifications for conserving genetic variation. Where breeding programs are designed to maximize short-term gains using high selection intensities and small effective sizes, semen or embryo banks are recommended to provide for ready access to alternative sets of alleles. Artificial Insemination is now used at least in cattle in most countries, so samples of many different strains could be collected with minimum cost. Storage of this semen (or of embryos) would be very cheap by comparison with the cost of maintaining discrete populations either on research farms or through restrictive legislation. Land (1981) has proposed the development of strains with divergent biological characteristics to increase future genetic flexibility. There will surely come a time when the question will be asked: "What is the number of breeds or strains which should, or even can, be maintained as separate entities within a country?"

REFERENCES

Becker, W.A. (1974). "Manual of Quantitative Genetics." 3rd Edition. Washington State University Press, Pullman.

Bichard, M., Pease, A.H.R., Swales, P.R., and Özjütük, K. (1973). *Anim. Prod. 17*, 215.

Bird, P.J.W.N., and Mitchell, G. (1980). *Anim. Breed. Abstr. 48*, 499.

Bohren, B.B. (1975). *Genetics 80*, 205.

Brascamp, E.S. (1973). *Z. Tierzücht. Züchtbiol. 90*, 1.

Comstock, R.E. (1977). *In* "Proceedings of the International Conference on Quantitative Genetics" (Pollak, E., Kempthorne, O., and Bailey, T.B., eds.), p. 705. Iowa State University Press, Ames.

Dempfle, L. (1975). *Genet. Res. 24*, 127.
Dickerson, G.E., and Hazel, L.N. (1944). *J. agric. Res. 69*, 459.
Eisen, E.J. (1980). *Z. Tierzücht. Züchtbiol. 97*, 305.
Enfield, F.D. (1977). *In* "Proceedings of the International Conference on Quantitative Genetics" (Pollak, E., Kempthorne, O., and Bailey, T.B., eds.), p. 177. Iowa State University Press, Ames.
Falconer, D.S. (1960). "Introduction to Quantitative Genetics." Oliver and Boyd, Edinburgh.
Falconer, D.S. (1977). *In* "Proceedings of the International Conference on Quantitative Genetics" (Pollak, E., Kempthorne, O., and Bailey, T.B., eds.), p. 101. Iowa State University Press, Ames.
Franklin, I.R. (1982). Chapter 10, this Volume.
Hammond, K. (1974). *In* "Proceedings 1st World Congress on Genetics Applied to Livestock, Madrid" Vol. 1. p. 425.
Harris, D.L. (1977). *In* "Proceedings of the International Conference on Quantitative Genetics" (Pollak, E., Kempthorne, O., and Bailey, T.B., eds.), p. 587. Iowa State University Press, Ames.
Henderson, C.R. (1977). *In* "Proceedings of the International Conference on Quantitative Genetics" (Pollak, E., Kempthorne, O., and Bailey, T.B., eds.), p. 615. Iowa State University Press, Ames.
Hill, W.G. (1974). *Animl Prod. 18*, 117.
Hill, W.G. (1977a). *In* "Proceedings of the International Conference on Quantitative Genetics" (Pollak, E., Kempthorne, O., and Bailey, T.C., eds.), p. 343. Iowa State University Press, Ames.
Hill, W.G. (1977b). *In* "Proceedings of the International Conference on Quantitative Genetics" (Pollak, E., Kempthorne, O., and Bailey, T.B., eds.), p. 367. Iowa State University Press, Ames.
Hill, W.G. (1982). *Proc. natn. Acad. Sci. U.S.A.* (in press).
Hinks, C.J.M. (1974). *Z. Tierzücht. Züchtbiol. 91*, 169.
Hopkins, I.R. (1978). *Anim. Prod. 26*, 267.
Hopkins, I.R., and James, J.W. (1977). *Anim. Prod. 25*, 111.
Hopkins, I.R., and James, J.W. (1979a). *Anim. Prod. 28*, 139.
Hopkins, I.R., and James, J.W. (1979b). *Anim. Prod. 28*, 65.
Hudson, G.F.S., Schaeffer, L.R., and Wilton, J.W. (1979). *J. Anim. Sci. 49*, 160.
James, J.W. (1962). *Heredity 17*, 487.
James, J.W. (1971). *Genet. Res. 16*, 241.
James, J.W. (1972). *Anim. Prod. 14*, 1.
James, J.W. (1978). *Acta. Agric. scand. 28*, 387.

James, J.W. (1979). *Livest. Prod. Sci. 6*, 215.
Jones, L.P., Frankham, R., and Barker, J.S.F. (1968). *Genet. Res. 12*, 249.
Land, R. (1981). *Livest. Prod. Sci. 8*, 95.
McClintock, A.E., and Taylor, J.F. (1982). Chapter 9, this Volume.
Nicholas, F.W., and Robertson, A. (1980). *Theor. Appl. Genet. 56*, 57.
Rendel, J.M., and Robertson, A. (1950). *J. Genet. 50*, 1.
Robertson, A. (1960). *Proc. R. Soc., Ser. B 153*, 234.
Robertson, A. (1961). *Genet. Res. 2*, 189.
Skjervold, H. (1963). *Acta. Agric. scand. 13*, 131.
Smith, C. (1969). *Anim. Prod. 11*, 433.
Smith, C. (1981). *Livest. Prod. Sci. 8*, 193.
Yoo, B.H. (1980). *Genet. Res. 35*, 1.

Discussion Summary

A. Comparison of Large and Small Breeding Programs

It is often assumed that larger dairy herd improvement programs (say 10^6 recorded cows) will be superior to smaller ones (say 10^3 cows) for both short-term and long-term rates of genetic improvement. Further, larger programs should produce cheaper semen through economies of scale. However, the predicted rate of genetic improvement in a large population may be reduced because of difficulties in controlling the important decisions concerning selection and utilization of superior animals. In addition, a breeding program may be small, yet still sell a large quantity of semen to farmers not participating in the breeding program. This allows the costs to be spread over many cows, just as in a large program.

If individual breeding programs are small, it is possible to have several instead of one large program in a country or region. This may provide added advantages in the utilization of epistatic interactions, overcoming possible genotype x environment interactions, and in gains from competition between the programs. However, competition between improvement programs in the market place also has potential detrimental effects. For instance, if imported semen which is well promoted but of doubtful genetic value is widely used, the effective number of cows available for selection and bull testing is decreased.

However, the number of cows recorded is not a complete description of the size of a program. The number on which there is sufficient information and which are actually available for selection may be much less than the total number of recorded cows due to periodic sale of herds, inconsistent use of AI or milk recording and regulations prohibiting the use of unregistered males through AI. A clear distinction must be made between the breeding program and the evaluation program. A co-ordinated evaluation program might provide information, particularly on breeding values, for several improvement schemes. This would allow some economies of scale, for instance in computing costs, but at the expense of a degree of self-sufficiency.

In group breeding programs in sheep and beef cattle, other considerations may favour a small sized program. Socio-political issues such as communication amongst group members, group identity, management by enthusiastic and committed farmers are important.

The effectiveness of genetic advice to the program directors may be related to program size. This advice may be better in small programs, either because geneticists have direct control of the program ("benevolent genetic dictatorship"), or because only enthusiastic and well-informed farmers are participating ("enlightened democracy").

It is not possible to make general recommendations on population size for all breeding programs. Each must be tailor-made to suit the circumstances in which it is to operate.

B. The Effectiveness of Very Intense Selection

Parent-offspring regression may be non-linear. For instance, a rare gene with a large effect would make extreme animals potentially very useful. Alternatively, these extreme animals could be the result of rare environmental effects and not at all useful. When considering extreme animals it must be remembered that a few individuals will be over three phenotypic standard deviations from the mean, even in a perfectly normally distributed population. Unfortunately, there is a lack of experimental data on the importance of non-linearity of the regression. In dairy cattle the dam-daughter regression appears to be linear for yield of milk and its components for the range $\mu \pm 3\sigma$ (μ and σ being the mean and phenotypic standard deviation of adjusted yield records expressed as deviations from appropriate herd-year averages).

Although non-linearity might not affect which animals are selected, it would alter the effect of important parameters such as selection intensity and population size.

C. Understanding the Nature of Quantitative Genetic Variation

Both extreme animals and the new variation generated during a selection program may be genetically unusual. If this variation were better understood, it could be utilized more effectively and it might be possible even to influence its generation.

Molecular genetics soon may be able to contribute to understanding in this area, for instance by distinguishing between structural genes and regulatory genes. However, animals which are phenotypically extreme are unlikely to be very different at a molecular level, because the co-ordination required between genes in development greatly constrains the variability that can occur, while avoiding lethality.

D. Planning for the Long-Term

There is a conflict in the design of breeding programs between maximizing short-term and long-term improvement. Producers of large animals cannot afford to plan on the basis of returns in 20 generations time, although this may not be the case for poultry breeders. The long-term advantages of large programs are difficult to predict due to: (i) the effects of mutation and migration, (ii) the long generation interval in cattle, (iii) the uncertainty of future breeding objectives, and (iv) the variability of appropriate discount rates to be applied when evaluating future gains. However, some traits such as milk production are likely to be desired for many years to come.

On the other hand, maximizing only short-term benefits could reduce the genetic variability available in the future. It is too costly to maintain populations of uneconomic animals, but embryos and gametes should certainly be stored. Plant breeders have already established genetic conservation programs and a 'Museum Bank' of *Bos taurus* semen exists in the U.K. In the future, genetic engineering techniques may be able to extract single desirable genes from completely unimproved populations or from other species and introduce them into commercial stocks, without also introducing many undesirable genes. This highlights the need for conservation of genetic material.

Part 8
Electronics

Implications of Developments in Engineering and Computing for the Genetic Improvement of Animals

Keith Hammond
A. E. McClintock[1]

Animal Genetics and Breeding Unit[2]
University of New England
Armidale, N.S.W.

Attention is drawn to the importance of recent developments in the computing and electronics industries and their impact on genetic improvement programs. The advantages of an electronic animal identification system (currently being developed) are listed. Developments in computer hardware and systems software are discussed in relation to strategies for implementing BLUP analyses. Several new measurement and mating procedures are proposed. These examples are used to demonstrate that the requirements of breeding programs for identification, measurement, recording and mating of stock could be made substantially easier, less expensive and more relevant but sufficiently reliable.

I Introduction

Once the basic theory of animal genetics was formalized, it was applied to many livestock populations in many countries but the rate of acceptance by individual farmers has varied greatly between populations. Among the reasons for this variation are the way particular farmers and industries view the impact of the modern breeding program on operations such

[1]*Supported by the Australian Meat Research Committee.*
[2]*A joint Unit of the New South Wales Department of Agriculture and the University of New England.*

FUTURE DEVELOPMENTS IN THE
GENETIC IMPROVEMENT OF ANIMALS
ISBN 0 12 078830 6

as stock identification and data recording, measurement of performance and mating. Of particular importance is the ease, cost, reliability and relevance of the intended changes to these operations, and existing traditions in the industry. Ease and cost are not always synonymous and both reliability of equipment and of results are important. The relevance of the intended changes must account for both the statistical and perceived associations between the operations and the breeding objective. An example of the importance of perceived relevance in the breeding operation could be the willingness of some registered dairy cattle breeders to sacrifice other production traits for longevity.

To maintain or increase the application of genetic principles, there is need for concerted effort to make the requirements of breeding programs in identification, measurement, recording, and mating of stock easier, less expensive, more relevant and sufficiently reliable. To an extent this is already happening. It is being encouraged by:

(1) The farmers' deteriorating terms of trade in many livestock industries, including the rapidly increasing costs associated with both the quality and quantity of labour.

(2) Recent increased research emphasis on defining breeding objectives and selection criteria for each production system. Undoubtedly, this has been encouraged by the increasing financial pressures in the industries but it is also being stimulated by consumer demand for better description of the product.

Further, there now exists substantial potential for the development of improved equipment because of:

(1) Progress in engineering hardware in areas other than animal science.

(2) The present and future possibilities for using microprocessors in many different ways.

Geneticists should not assume a wait-and-see attitude for the development of new technology which would assist the application of genetic principles. The situation is ripe for instigating multidisciplinary R & D efforts involving groups of engineers and of computer system specialists. Similar arguments to those which have been put for multidisciplinary efforts to determine breeding objectives and selection criteria are also relevant here.

II Production Operations and Application

A. *Identification and Data Systems*

Within a few years cheap and permanent (life-time) electronic systems for the identification of individual animals will be available to the livestock industries. The capabilities of these systems for dairy cattle were described by Hooven (1978). Hammond *et al.* (1979) specified the general requirements for electronic identification systems for use under extensive livestock management. Briefly, these systems will be capable of:

(1) Identifying individual animals within a group at distances up to 30 m.
(2) Recording and storing data on large numbers of animals.
(3) Being interfaced with various ancillary devices, e.g., electronic scales and other electronic measuring instruments, electronic drafting gates and other computers, directly or via acoustic couplers.
(4) Monitoring body temperature, e.g., for use in oestrus detection.
(5) Being used in the paddock, the yards and the office for making selection decisions.
(6) Identifying animals at abattoirs to enable classification data on individual carcasses to be returned to the breeder.

The information collected could be processed on the farm and through Networking systems. This would help overcome the 'turnaround' problem of many of the current computing configurations in the livestock industries where turnaround with a central processor is via the postal system. Such Networks would provide data security and also allow data to be pooled across herds for the estimation of genetic parameters and, where relevant, of breeding values.

Before electronic systems of animal identification are introduced it will be desirable to define national codes for appropriately identifying all stock, the one code covering all on-farm and off-farm operations. Holm (1981) has described a code proposed for the U.S. livestock industries. Further, much work remains to be done in defining the precise data systems best suited to an industry. A start has been made on the development of such systems for dairy cattle (e.g., Speichter, 1981).

The capacity limitations of main frame computers when handling such problems as large mixed model analyses are rapidly being reduced. For example, most Best Linear Unbiased Prediction (BLUP) models in practical use have relied on the iterative solution of systems of equations with large, sparse coefficient matrices. The usual 'Cornell' strategy is to read and process one equation at a time from backing-store (disc) into fast-store (core or semiconductor memory). This procedure is not limited in practice by the size of fast-store (but it can be limited by the speed of transfer of an equation from disc to core unless low level languages are used). Two developments which have taken place since this strategy was first used are the increase in the maximum array size which main frame computers can address and the 'Distributed Array Processor' (DAP).

The increase in maximum array size can allow larger matrices to be manipulated in core. Not only has the price of core fallen considerably, the appearance of time sharing systems which automatically move arrays and program between core and disc has simplified the programmer's task. Thus, programs can be written as though very large matrices were being held in core. However, care has to be taken not to cause 'thrashing', which can occur when a disc-to-core transfer is generated every time an array element is accessed. For example, with DEC or ICL FORTRAN, two-dimensional arrays are stored column-by-column.

Distributed Array Processors (Gostick, 1979) are now being used for Econometrics and for Weather Research which involve the solution of large sets of equations. Essentially the method involves a simultaneous processing of sections of an array by perhaps 1000 to 16000 micro-processors. Although expensive at present and apparently unique to ICL, this technology could eventually ease some of the present BLUP computing problems.

Micro-computers can already be made to handle at least some BLUP models. Admittedly they will be slower but, with sufficient disc space, a current Apple II_{TM} could be used to complete a BLUP analysis of preadjusted data involving several hundred sire equations and incorporating the Numerator Relationship Matrix.

B. Measurement

The 'measurement paradox' is that recommendations are being made to measure more and more traits in animal breeding programs whilst calls are also being made to decrease the total cost of measurement. The dairy industry has overcome this paradox at least once when, during the 70's, butterfat testing of milk was moved off the farm to central laboratories where newly developed equipment permitted protein measurements, cell counts (for mastitis) and pregnancy tests to be made on milk samples from individual cows. This was done at a much lower cost to the industry relative to the cost of the original labour-intensive milk recording system. In many countries this change was made in the brink of time.

Automation must be emphasized when developing measuring equipment for use in the livestock industry. Once electronic identification becomes a reality the automatic measuring of some traits will even be possible whilst commercial animals are grazing in the paddock, e.g., body weight and temperature as the animals walk through a race to water.

In designing a breeding program we aim to maximize the correlation between the breeding objective and the selection criterion. To do this we require good estimates of the necessary genetic parameters for all economically important traits. As the magnitude of parameters such as heritabilities decrease very large numbers of observations are necessary to obtain useful estimates. In the past we have tended to downgrade some of these lowly heritable characters, yet some are economically very important. In addition, current methods for obtaining information on individual animals for some apparently important traits may be slow and laborious. Intake, feed requirements for production, body composition and many components of reproduction fall into this category. Further, proposals have been made to describe the nature of the variation for individual components of production systems (e.g., Land's, 1981, physiological criteria) or of product (e.g., the individual milk fractions of Everett, Hammond and Taylor, 1982).

Procedures and equipment which would enable large numbers of observations on these and other supposedly important traits would be of great value in our efforts to define 'efficient production' in different environments. This statement seems obvious but again is meant to encourage greater emphasis by researchers to identify the variables and develop the equipment for their rapid measurement. The advent of the micro-processor should assist in moving away from some of the more laborious procedures in current use. The recent development of computerized tomography for estimation of body composition on the live animal (Skjervold *et al.*, 1981) is a good example, with the potential to provide a quantum jump in our understanding of the genetics of the growth processes in different environments.

Genetic improvement involves manipulating the current biological materials. We have tended to concentrate on accumulating small changes. Nevertheless, if we are in a position to measure the appropriate variables on very large numbers of animals we can then capitalize on the genetic outliers. Possibly, this becomes even more important now that the transfer of DNA between species is imminent. However, in the short-term it may be more practical, for example, to capitalize on outliers by utilizing both overt and covert (Rendel, 1977) variation for particular characters. Again, this will require taking a larger number of observations for the critical variables. What are the critical variables, and are we in a position to identify outliers which would assist us in achieving the following biological models:

High fecundity with easy-care for particular environments.
Beef cows capable of two gestations per year.
Dairy cows capable of maintaining peak lactations for several years.
The ability to tune metabolism to seasonal conditions.

Many of the past evaluations of measuring equipment for use in performance recording have examined its precision and accuracy using statistical tests such as described by Grubbs (1973). The bias of a particular instrument is of no importance since animals to be compared will usually be measured using the same instrument. Genetic parameters will therefore be independent of instrument bias. Even relatively large random errors associated with an instrument could be tolerated. For example, the heritability of the trait as measured might still be large enough to be worth selecting for (Returns - Cost would be a good criterion to judge equipment). If this were done more frequently it may prevent the down-grading of some inexpensive and easily applied procedures.

C. Mating

Mating is an important operation in the breeding program and, compared with the herd or flock not conducting a breeding program, involves added difficulties. For example, ensuring randomness, allowing for collection of pedigree information and facilitating direct comparisons across groups. Where artificial insemination replaces natural mating there are further difficulties with oestrus detection and insemination.

These problems could at least partially be overcome by the development of a semi-automatic AI procedure. For example, a package of semen could be positioned in the reproductive tract of all females to be mated; the semen could be released in a particular female at ovulation. Alternatively, for extensive management systems the combination of electronic identification and temperature monitoring with a 'catching device' would at least collect animals to be inseminated. It is possible that a hormone 'sniffer' (similar in principle to a smoke detector) could be used to draft oestrus females even if they were not carrying an electronic identification device. Mechanical pedometers have been found to be a useful way of detecting oestrus cows; an electronic equivalent could supplement temperature data acquired by an implanted electronic identification device.

III Conclusions

In this short communication we have endeavoured to show by example that the time is ripe for geneticists to encourage the development of equipment which would in turn assist with the better definition of selection criteria.

REFERENCES

Everett, R.W., Taylor, J.F., and Hammond, K. (1982). *J. Dairy Sci.* (in press).
Gostick, R.W. (1979). *ICL Tech. J.* May, 116.
Grubbs, F.E. (1973). *Technometrics 15,* 53.
Hammond, K., Davies, L.W., Fletcher, N.H., McIntosh, A.I., Webb, R.F., and Yerbury, M.J. (1979). Proc. Inaugural Conference of the Australian Association of Animal Breeding and Genetics, Armidale. p. 100.
Holm, D.M. (1981). *J. Anim. Sci. 3,* 524.
Hooven, N.W. (1978). *J. Dairy Sci. 61,* 1167.
Land, R.B. (1981). *Livest. Prod. Sci. 8,* 203.
Rendel, J.M. (1977). *In* "Proceedings of the International Conference on Quantitative Genetics" (Pollak, E., Kempthorne, O., and Bailey, T.B., eds.), p. 23. Iowa State University Press, Ames.
Skjervold, H., Grønseth, K., Vangen, O., and Evensen, A. (1981). *Z. Tierzücht. Züchtbiol. 98,* 77.
Speichter, J.A. (1981). *J. Anim. Sci. 53,* 531.

Discussion Summary

The creation of national data files would be of advantage: (1) where genetic links exist between herds, thus allowing across-herd rankings, or (2) where there is difficulty in obtaining national statistics in other ways, e.g., surveys. Such national statistics could permit a forecast of production levels to enable the formulation of better marketing and processing decisions.

If such national data files are to be created, consideration will need to be given to: (1) animal identification, and (2) storage and communication of data.

A national numbering system would be of value not only for on-farm performance recording, but also use by veterinarians and abattoirs, to allow automatic data capture of herd (flock) health and carcass information. The reuse of identification numbers would cause problems, and each number would have to be unique.

Various possibilities for the storage and communication of performance data were discussed. Farmers could transmit data via telephone lines, into a computer possibly operated by either a local stock agent or veterinary surgeon. An increasing number of farmers undoubtedly will obtain their own microcomputers, together with floppy discs for data storage. However, the unreliability of floppy discs might increase the desirability of using a central computer linked via telephone lines, as an additional data storage device. If a central computing facility were given the task of collating data from a large number of on-farm files, there could be considerable problems of compatability of record layout and the coding of information. To avoid this proliferation of incompatible files at the farm level, the N.Z. Dairy Board intends to create a single national database to which a producer would have appropriate access via telephone lines and an on-farm terminal (or microcomputer).

As farmers would use the computer mainly for management decisions, it would be necessary to convince them of the importance of making records available to a central data bank.

Potential changes in the production environment due to engineering and electronics were considered. For example, the development of an automatic milking machine to which the cow presented herself at her own convenience might permit genetically superior cows to better express their milking potential.

Index

2 3 4 5 6 7 8 9 0 1
A B C D E F G H I J